# Natural Kinds and Genesis

# Natural Kinds and Genesis

## *The Classification of Material Entities*

Stewart Umphrey

LEXINGTON BOOKS
Lanham • Boulder • New York • London

Published by Lexington Books
An imprint of The Rowman & Littlefield Publishing Group, Inc.
4501 Forbes Boulevard, Suite 200, Lanham, Maryland 20706
www.rowman.com

Unit A, Whitacre Mews, 26-34 Stannary Street, London SE11 4AB



British Library Cataloguing in Publication Information Available

**Library of Congress Cataloging-in-Publication Data**

Names: Umphrey, Stewart, 1942– author.
Title: Natural kinds and Genesis : the classification of material entities / Stewart Umphrey.
Description: Lanham : Lexington Books, 2016. | Includes bibliographical references and index.
Identifiers: LCCN 2016025725 (print) | LCCN 2016028072 (ebook) | ISBN 9781498531412 (cloth : alk. paper) | ISBN 9781498531429 (electronic)
Subjects: LCSH: Entity (Philosophy) | Categories (Philosophy) | Classification. | Naturalism. | Monism. | Classification of sciences.
Classification: LCC BD336.U47 2016 (print) | LCC BD336 (ebook) | DDC 111--dc23
LC record available at https://lccn.loc.gov/2016025725

∞™ The paper used in this publication meets the minimum requirements of American National Standard for Information Sciences Permanence of Paper for Printed Library Materials, ANSI/NISO Z39.48-1992.

Printed in the United States of America

Seth Benardete

1930–2001

Il ne regardait que la salle. Il aspirait la grande bouffée brûlante,
au bord du trou.

# Contents

# Acknowledgments

It was my late colleague and friend, Bill Edgar, who, during a year-long seminar we held on universals and particulars, first brought home to me the need to think more deeply about natural kinds. Without his prompting this book would probably not have been written.

David McNeill read a draft of early chapters and offered very helpful advice. Dylan Casey, Nick Maistrellis, Susan Paalman, Kathy Blits, Chester Burke, and Michael Comenetz all came to my aid while I was preparing to write chapters 6 and 7. Alan Pichanick helped me to see more clearly a way in which natural philosophy today can be rejuvenated. To all, my thanks.

I am grateful also to St. John's College for allowing me to learn by teaching in its laboratory program, and to the National Endowment for the Humanities for a grant (2001–2003) that enabled me to study topics in evolutionary theory and molecular genetics.

# ONE
# Questions

In an essay on language and reality, Jorge Luis Borges refers in passing to an old Chinese encyclopedia in which animals are divided into the following kinds: those belonging to the Emperor, embalmed ones, those which have been trained, suckling pigs, mermaids, fabulous ones, stray dogs, those included in this classification, those trembling as if they were mad, innumerable ones, those drawn with a fine camel's hair brush, others, those having just broken a flower vase, and those resembling flies from a distance.[1] We laugh. Why? Because it is patently absurd: as a classificatory scheme it fails to distinguish what is real from what is merely subjective or conventional, while as a list its omissions, redundancies, and general hodge-podgery make it worthy of Rabelais at his best. It shows what can happen when taxonomy goes on a holiday.

What attitude does our laughter express? Complacency, perhaps; we may reject out of hand this strange way of classifying animals and rest content with our own, in which case our laughter is Hobbesian, displaying sudden glory before the inferiority of another. Or it may instead express liberation from the conviction that our own classificatory scheme is correct, in which case Borges' essay works like those jokes that crack us up by exposing the folly of conventional wisdom. It may occur to us, if only for a moment, that our preferred scheme is likewise ridiculous, that the joke is really on us. We may begin to suspect that all such schemes are provisional at best, that they never measure up to the way things are.

Cracking up may fill one with angst. Reading Borges' essay caused Michel Foucault to wake up laughing from a dogmatic slumber, he said, and yet so shattered by it was his way of ordering things that the very distinction between same and other evanesced.[2] Yet these moments of liberation need not be so world-shattering. We may instead wonder whether others would find our classificatory scheme as strange as we find theirs, and ask ourselves what the true ordering of things is and how one might go about discovering it.[3] Our own scheme has been defamiliarized, then, without being destroyed. Held less tightly by beliefs we hold, we are free to inquire without having lost our bearings altogether; a former conviction has given way not to utter bewilderment, or to the contrary conviction, but to an attitude of entertainment in which we can examine the different ways in which people make sense of things, in order to discover (if we can) the way things are in reality or by nature.[4] This is how natural philosophy begins.

Though he may provoke such questions, Borges does not create them out of thin air. Nor must we be worked on by devices so artfully contrived or by interrogators so philosophically astute to wonder about such things; for even when inattentive to the possibility that our worldviews are cognitively inadequate, as most of us are most of the time, we are predisposed to admit as much. To see that this is so, consider what we call "common sense." It assures us that we live amid myriad things, some of which are manmade or conventional, others natural. Things natural, though variously various, fall for the most part into discrete kinds—animal, vegetable, mineral, and so forth—which become evident to us as we become receptive to them. Yet no less commonsensical is it to think that people sometimes regard as real things that are not: Olympian gods, unicorns, elves, ghosts, Santa Claus, the Bogeyman, to mention but a few. We are not surprised, then, to learn that our ancestors regarded as natural some kinds we regard as arbitrary or imaginary; and that our descendents will think the same about us is an inference we have little reason to reject. It is commonsensical also to acknowledge the existence of kinds that do not wear their reality or unreality on their sleeves. Take for example the class of drugs, or weeds, or beasts, or the division of mushrooms into edible and poisonous varieties: are these kinds natural even though they depend partly on our concerns? Or take the various abnormalities people may suffer. How abnormal a state can one be in and still be human? What, if anything, do natural and normal have to do with each other? The very distinction between natural and conventional is sometimes hard to make, and denying the existence of these hard cases, or drawing bright lines about the grey areas they represent, would be paradoxical.

Being of two minds about the reality of apparent kinds is particularly manifest right here in the United States. On the one hand, sharp divisions along racial, sexual, ethnic, and religious lines are as American as apple pie. Moreover, like people everywhere, we recognize one another in terms of social types. Do I sell used cars for a living? Then I am a used-car salesman. Do I say little and keep to myself? Then I am a loner, the silent type. It seems too that I have grown into certain roles and routines; they fit me as well as my given name. Were the objectivity of such types to vanish, the reality of everyday life would be destroyed.[5] America presents no exception to the proposition that man is a classifying animal, with a tendency to take for granted the results of its sortal activities. On the other hand, we deeply resent being stereotyped by the pigeonholing proclivities of others, on the ground that no such typecasting can capture the real you or me.[6] We resist efforts to make us conform to roles and routines which keep us from becoming the unique individuals we are. Idiosyncrasy may be frowned upon, yet it points to what is real in each of us. Play-acting among adults is discouraged, yet it helps bring to light the theatrical character of everyday life and fosters our ability to exemplify social types in the manner of actors exchanging masks. Individualism disposes us to regard such types as unreal.[7] And since America is a rather open society, this tendency flourishes as we learn about alternative classificatory schemes accepted by people in other cultures or by experts in our own (according to whom whales are not fish, electrons are not animal or vegetable or mineral, and so on). We easily infer that all kinds commonly regarded as real are in fact ideal types projected onto the world like labels affixed to things by persons largely unknown, operating behind our backs, for reasons of their own. Nominalism is as American as apple pie.

It seems natural, then, to wonder whether the kinds we ordinarily take to be natural are natural in truth. All we need is a catalytic circumstance to get us started, and these occur everywhere; they include not just Borgesian or Platonic artifacts, but everyday uncertainties of the sort just mentioned. And when we begin to examine them our questions multiply fruitfully. Soon we ask whether there are any *natural* kinds at all. Any *what*? What is it to be a natural kind? This question must be answered before the others can be answered well. For us, then, the proper order of inquiry is to ask first, "What is a natural kind?" and then, "Which, if any, are the natural kinds?" These two questions have put us on a path in natural philosophy. While proceeding on this path we may find that answering the first must involve trying to answer the second as well, but if so we should not let this finding obscure their proper order.

One may think that our inquiry will prove idle. Perhaps so, but one should not presume as much on the grounds that our questions must not have arisen in everyday life; for again, we are all pre-philosophically open to the possibility that some given types are unreal, indeed to the possibility that all classificatory schemes are merely provisional, and it would be dogmatic to presume that all attempts to answer these questions are doomed to failure. Why then do philosophical inquiries seem so often to come out of the blue? Why is it commonsensical to regard with suspicion the question, "What is a natural kind?" The following analogy may help.

People everywhere are more or less open to the possibility that the letter of the law falls short of its spirit, that the society in which they live is not perfectly just. There are circumstances in which they question whether their society *is* just, and this question, once raised, begets others;

one begins to wonder whether any political society is truly just, and what it is to be just. Ordinarily, of course, it seems neither necessary nor right to stir up such questions, so when they do arise they are not *our* questions; they come out of nowhere, and those prompting us to answer them resemble uninvited guests bearing unwanted gifts. We are apt to take offense. Yet there are extraordinary circumstances in which raising and pursuing such questions seems fitting, even urgent, as in times of political upheaval or on those occasions in the lives of each when we confront morally hard choices and wonder what the right course of action might be. Unsettling circumstances of this sort may not be the norm, yet there is no human society in which they do not sometimes occur and it is in disruptions of this sort that moral or political philosophy first takes hold. Similarly, it is in disruptions of the sort I have described that our questions about natural kinds first arise. Exploiting a distinction made by Thomas Kuhn, we may say that it is in revolutionary rather than normal situations (of one sort or another) that philosophy (of one sort or another) enters our lives.[8]

One may still regard our present inquiry as frivolous, on the grounds that human cares are practical rather than theoretical, that moral and political issues are more important than metaphysical conundrums, and consequently, whereas the question "What is justice?" is one we should sometimes take seriously, the question "What is a natural kind?" is not. Our inquiry is plainly theoretical rather than practical. But issues concerning natural kinds are not merely theoretical: consider for example the question whether our familiar divisions along racial, sexual, or ethnic lines have any basis in nature, an issue now hotly debated. Nor do fundamental questions about justice arise only in times of personal or political difficulty: consider for example what happens when we read Aesop's fables, *Gulliver's Travels*, Plato's *Republic*, or the Bible.[9] I agree that our inquiry, unlike those undertaken by some in times of moral or political upheaval, carries little sense of urgency, and that matters claiming our immediate attention can seem very important. But the important is no more reducible to the urgent than the good is to the necessary. In natural philosophy, time is rarely of the essence. Its practitioners can and should therefore avoid any rush to judgment. Indeed, being circumspect and tentative seems apt, given the obscurity of the issues before them and the theoretical importance of getting things right. Natural philosophy has about it an air of leisurely consideration, not to say idle speculation. Finally, I agree that most people do not put our questions high on their list of cares: they are otherwise engaged.[10] It is no great surprise then that philosophers are as rare as theoretical physicists, or that people tend to regard the unusual as unnatural. From this, however, it does not follow that our inquiry is frivolous or that natural philosophy has no roots in the everyday world of gardeners, bird watchers, hunters, metal workers, and housewives.[11]

One may still think that our first question has been answered by those philosophers in the modern "tradition of natural kinds" who claim that a natural kind is:

1. a class that exists in nature independently of whatever we may think or say about it;
2. whose members are such things as cats or samples of gold;
3. which share essentially and distinctively some property or small set of properties; and
4. that serve to explain their other properties and behavior according to physical laws or underlying causal mechanisms.[12]

Some such account is accepted by most natural-kinds realists today. Yet we should now refrain from accepting it, for two reasons. First, it combines common opinion, all decked out in the technical language of conditions 1–3, with some cutting-edge scientific theory indicated by condition 4, and this is a dubious marriage. I shall undertake to show as much in chapters 6 and 7. For now, though, it suffices to note the following.

Ordinarily we use what people do to determine who they are, and who they are to determine what we can expect from them; and in making these inductions we resort to commonly presupposed types. ("Why are you acting this way?" "We are Americans!" "Ah, that explains a lot.") These types are generally of two sorts: of the *person* or *thing* that tends to behave thus and so (the teacher, the hammer, the lion, gold), or of the *behavior* usually displayed by such a person or thing in specific circumstances (teaching, hammering, predation, malleation).[13] Behavior-types are generally prior in the order of knowing, person- or thing-types in the order of being, and yet so intimately related are they in everyday thought that we can hardly tell them apart. This intimacy, more often felt than noticed, licenses our easy inference from what you do to who you are and from who you are to what we can expect from you. Indeed, we commonly understand the *being* of things in terms of what they *do* when they are doing their own thing. The concept of a natural kind has its native home here.

Scientific differs from everyday explanation in several important respects. First, it it takes causal relations to be necessary—when, at any rate, the concept of causality applies—and all such relations in nature are describable by fixed laws. Second, it rejects teleology: the biotic systems we call "teachers" do not move about for the sake of educating others and making a living. Third, it tends not to acknowledge things as we commonly experience them, but instead such objects as quarks and leptons, molecules and alleles, which we never directly experience. These objects scientists regard as instantiations of types, but do they regard them as natural kinds? Are they not instead heuristic or mnemonic devices? In fact, as science has progressed, more and more natural-kind terms have fallen away into the conceptual wastebasket of history, which suggests that in the final theory of everything there will be no room for natural kinds.[14] Talk of them may still have some pragmatic value. One may still hold that conditions 1–4 jointly characterize what it is to be a natural kind. But to go on thinking that the concept must have some theoretical use would be like holding onto the Euclidean definition of a square in non-Euclidean geometry: possible but weird. Is it not reasonable to suspect that marrying conditions 1–3 with condition 4, as natural-kinds realists today understandably wish to do, is like trying to mate a human being with a very advanced robot?

There is another reason for bracketing, as it were, this initial answer to our question: it is not yet *our* answer; we have not done the hard work of discovering it for ourselves. Hence, for us, it is but one relevant opinion among others, all in need of further examination. These opinions have roughly three sources: common sense, scientific theory, and the philosophical tradition. For us as natural philosophers, none of these resources is simply authoritative, yet each seems to provide more or less reliable heuristic wherewithal; each is to be taken seriously as a locus for philosophical investigation. Although ideally we would make use of all three resources, practical considerations prevent me from using the third as well as the first two in this book. For an account of what one finds by investigating natural kinds under the guidance of relevant philosophical traditions, readers are referred to my forthcoming supplementary volume entitled *The Aristotelian Tradition of Natural Kinds and Its Demise.*

The concept of a natural kind has its native home in our pre-scientific understanding of things—one sign of which is the fact that both great traditions of natural kinds, the Aristotelian in premodern times and the Millian in our era, remain fairly close to everyday experience, or to Kripke calls "our intuitions." In other words, it is by investigating what Husserl calls the *Lebenswelt* (life-world) that one first comes to understand the natural order in terms of natural kinds.[15] Our initial questions about such kinds have the same home. Accordingly, I proceed first under the guidance of common sense or the folk naturalism of everyday life (in chapter 2–5), and only then under the guidance of contemporary scientific theories (in chapters 6–7); and I do so always with a view to answering two main questions, one conceptual and the other empirical, namely, "What is it to be a natural kind?" and "Are there in fact any such kinds?" The following is a preview of coming accounts.

### *Chapter 2*

Our pre-scientific understanding of natural things presupposes that their reality does not depend on how we regard them, and that the way we ordinarily regard them is heuristically, if not cognitively, valuable. There is today a kind of philosophical antirealism that rejects the first of these presuppositions, and a kind of philosophical naturalism which rejects the second. Both would object to my way of investigating natural kinds. I defend natural philosophy, as I conceive it, against these two objections.

*Chapter 3*

Natural kinds as types seem to be universals, which is why the reality or unreality of universals was a central issue in the Aristotelian tradition of natural kinds, and why it must be an issue for us as well in our present investigation. I argue for realism with respect to universals. Yet the reality of universals does not entail the reality of natural-kind types; and to ascertain whether there are any types of this sort, one must proceed naturalistically and not simply in a broadly logical manner. I argue, moreover, that one proceeds naturalistically in this case only if one bears firmly in mind that the tokens of exemplary natural-kind types are entities which remain the same in and through change. Such entities philosophers call "endurants" or "continuants." Continuants are essentially beings-in-becoming. Thus natural kinds cannot be understood apart from *genesis* or from the ever-changing material world as such. The investigation of continuants, what they are and whether they exist, is broadly metaphysical. So too, therefore, is the investigation of natural kinds, what they are and whether they exist. In the next two chapters I shall proceed under the guidance of what philosophers today call "folk metaphysics," and by so doing reconnect what they have tended to take apart and treat separately: the metaphysics of natural kinds (as types or classes) and the metaphysics of continuants (as enduring individuals or substances).

*Chapter 4*

Exemplary continuants for us include substances such as water and gold, and individuals such as oak trees and squirrels. Taking a putative continuant of the last-mentioned sort as representative, I first examine the metaphysical difficulties surrounding it, with a view to sharpening the notion of a continuant and clarifying the sense in which any entity of this sort has an essence even though it depends, necessarily, on the ever-changing material world to which it belongs. Then I examine the epistemological difficulties surrounding such entities, and try to answer the question whether material continuants exist. On reflection it appears reasonable to continue supposing that they do, even if many of the things commonsensically thought to to be continuants are not individual entities at all. I accept continuant realism as a reasonable hypothesis.

*Chapter 5*

Therefore, in the primary sense of the term, a natural kind (as a class) is one whose members are those continuants which exactly or very closely resemble one another in virtue of their essences or natures. Alternatively, it is all and only those continuants which exemplify a specific natural kind (as a type) in virtue of their essences or natures. This answers the first of our two leading questions, namely, "What is it to be a natural kind?" I defend this answer against several possible objections, including the claim that there is no such thing as *the* concept of a natural kind, or that this concept has no distinctive role to play in our thought. Then, having distinguished and appraised the varieties of natural-kinds realism and antirealism, I argue that common sense alone does not enable one to ascertain whether there are in fact any natural kinds. To answer this second, more empirical question we need to enlist the help of natural science.

*Chapter 6*

Adapting a Putnamesque term to my present purposes, I call *being hydrogen oxide* (or $H_2O$), for example, a "scientific stereotype" and call "stereotokens" whatever scientists take to instantiate this type. The question, then, is whether these stereotokens are genuine continuants and whether the stereotypes they instantiate, or the specific classes to which they belong in virtue of their natures, are indeed natural kinds. I begin by considering the kinds posited in theoretical chemistry and particle physics. It does appear that some chemical kinds, at least, are natural kinds in the primary sense of the term. Yet even here we need to proceed cautiously, since modern science is inhospitable to natural-kinds realism in two related ways: first, it tends to reduce apparent continuants to objects that are not individuals at all; second, it tends to replace the more commonsensical view of nature, in which localized natural kinds are central, with one in which comprehensive laws of nature are central. These two tendencies, taken together, strongly suggest that natural-kinds *anti*realism is true. In their eagerness to identify the kinds posited in chemistry or physics with natural kinds, scientific essentialists have paid too little attention to this fact.

*Chapter 7*

I first conduct a similar examination of biological stereotypes and stereotokens, and then argue for a narrowly delimited natural-kinds realism. Although biological species are still sometimes taken to be natural kinds, those posited in evolutionary theory are genealogical ($species_g$) rather than typological ($species_t$); and although $species_g$ as units of evolution are indeed stereotokens, they are not genuine continuants. Codons are among the stereotokens posited in molecular genetics, but they cannot be genuine continuants, since as functional units they depend essentially on the organisms to which they belong. And as for the organisms themselves, biological theory today gives mixed signals: so deep is its metabolic, compositional, and ecological involvement in the world that biologists tend to regard a living thing as many rather than one, but they also tend to agree that the evolution of organic individuals has occurred. So perhaps some living things are genuine continuants; and perhaps some cases of paraphyly and polyploidy do give us reason to think that a few biological kinds are natural kinds in the primary sense. The exemplary natural kinds, however, appear to be chemical. Their members appear to be continuants, and the classes they constitute appear to be well defined in terms of their natures. Scientific theory teaches us that all such stereotokens are emergent entities, and that such emergence correlates strongly with symmetry-breaking events. Accordingly, to defend natural-kinds realism and thwart the reductionist tendencies in natural science, essentialists today must show that substantive emergence (*genesis*) has in fact occurred. This I undertake to do.

*Epilogue*

By recourse to common sense and scientific theory, we have discovered what it is to be a natural kind and found reason to believe that such kinds exist. Recourse to relevant philosophical traditions, our third heuristic resource, is beyond the scope of this book, and yet some anticipation of what one would find by proceeding under their guidance is not out of place. First, apart from Aristotelians and Millians (those belonging to what Ian Hacking has identified as the modern tradition of natural kinds), natural philosophers have tended to be antirealists with respect to natural kinds, and many have been antirealists with respect to individual continuants as well. Second, in the principal scientific revolution for us, the Aristotelian eido-centric conception of nature was replaced by the Galilean nomo-centric conception. These divergent ways of conceptualizing the natural order are not obviously complementary. Newton, for one, took them to be mutually incompatible. Third, natural philosophy has always been subject to crisis; and the fourth major crisis, brought about by Kant's "critical" philosophy, is still with us. Natural-kinds realists and antirealists would do well to make themselves fully cognizant of these facts, not because they prove decisive in the realism-antirealism debate—they do not—but because they enable one to become still better acquainted with the problem of natural kinds as a problem; and this after all is progress of a sort. It is in fact the sort of progress one grows to expect in philosophy.

This book is intended primarily for those with some training in academic philosophy, who are interested in the metaphysics of natural kinds and their individual members. If they do not also have much expertise in the philosophy of chemistry, physics, and biology, they may find parts of chapters 6 and 7 rather hard to understand. I have tried to make the contents of these chapters as accessible to them as I can. Those, on the other hand, who do have some such expertise may find parts of chapters 6 or 7 to be less in-depth than they would like. I have tried to make as clear as I can how these chapters, just as they are, fit into the argument of the book as a whole.

Readers of both sorts will notice that although I count myself among those renewing natural philosophy today, this book is short on doctrine. It does not, for example, present anything like a complete theory of natural kinds. Yet I do clarify the concept of a natural kind and give reasons for believing that natural kinds exist. Moreover, by reaching these conclusions in a somewhat roundabout way, I have drawn (as it were) a map of the rugged conceptual landscape on which all natural philosophers today must proceed whether they like it or not. This map is necessarily incomplete, inasmuch as my focus is on natural kinds and not, for example, the nature of causality or of spacetime. It may also be in need of revision even with respect to exemplary natural kinds and the continuants which are their members. It is thus but a draft of a map, offered here so that others may use it as a guide in their own efforts to understand natural kinds and to draw a better map.

## NOTES

1. J. L. Borges, "The Analytical Language of John Wilkins," in *Other Inquisitions 1937–1952*, tr. R. L. C. Simms (New York: Washington Square Pr., 1965), 108. Did Borges invent this entry? If so, he may have been inspired by the lists in Sei Shônagon's *Pillow Book*.
2. M. Foucault, *The Order of Things* (New York: Vintage Books, 1973), xv.
3. G. Lakoff 1987, 92ff.
4. I am not now questioning the supposition that every natural kind is real and every real kind natural. The first conjunct will be affirmed in chapter 2. Reasons for denying the second will be offered in chapter 3.
5. On the typicality of things given in everyday experience, see: Husserl 1973, 36–39, 334–35; Gurwitsch 1964, 238–45; Schütz 1967a, 59–60, 73–74, 281–86, 348–52, and his 1967b, 176ff. The constructivist character of Schütz's account is announced in the original German title of the last-mentioned book: *Der sinnhafte Aufbau der socialen Welt*.
6. On the anonymity that comes with "typification," see cf. P. L. Berger and T. Luckmann, *The Social Construction of Reality* (Garden City, NY: Doubleday Anchor, 1967), 31–34, 39, 68, 142. Consider also Plato, *Parmenides* 131b7–9.
7. On our individualism, see cf. A. de Tocqueville, *Democracy in America*, tr. G. Lawrence (Garden City, NY: Doubleday Anchor, 1969), 435–36, 506–09. For different elaborations of this fact, see cf. E. Goffman, *The Presentation of the Self in Everyday Life* (Garden City, NY: Doubleday Anchor, 1959) and *Interaction Ritual* (Garden City: Doubleday Anchor, 1967); R. F. Baumeister, *Identity: Cultural Change and the Struggle for Self* (New York: Oxford University Press, 1986), 76–95.

8. Kuhn 1970, passim. What Kuhn calls "revolutionary science" is hardly distinguishable from natural philosophy as traditionally understood. Galileo and Newton regarded themselves as natural philosophers.

9. The difference in question here is not due simply to the fact that the concept of justice is evidently normative whereas the concept of a natural kind is not.

10. This is particularly true in the United States, where citizens have long been trained to value labor over leisure and to scoff at anything that sounds very theoretical.

11. Some philosophers believe that reflection of any sort removes one from everyday life. I admit a distinction between engagement and observation, dwelling-in and dwelling-on, and agree that "pre-reflective experience" is in some sense the default position. But I fail to see how such experience alone could constitute a human form of life. Is the now-fashionable idea of immaculate dwelling-in any more tenable than the once-fashionable idea of immaculate perception?

12. Hacking 1991, 110ff. The definition offered here is rather "Peircean": Hacking, op. cit., 120 with Peirce 1940, 600f. Peircean conceptions of natural kinds now prevail among natural-kinds theorists; see chapter 5 of this volume.

13. I extend to nonhuman things a distinction made by Schütz in his 1967b, 187f.

14. Cf. Russell 1948, 461f, and Quine 1969, 138—both of whom fall within the tradition Hacking describes. One is reminded of Wilfred Sellars' famous distinction between the manifest and the scientific "images." Natural kinds belong to the former, not obviously to the latter as well.

15. Compare the laws-of-nature tradition instituted by Galileo, Bacon, Descartes, and others, which depends partly on a critique of this "unscientific" approach.

# TWO

# Back to Nature

Some philosophers may have been free to investigate nature directly, without any necessary preliminaries, but most have not. Democritus and Empedocles had to deal with the Eleatic critique, according to which things natural only seem to exist. Aristotle had to deal with the Socratic-Platonic turn away from the natural world to human affairs and timeless beings. Seventeenth-century natural philosophers had to deal with the authority of Church doctrine, according to which the natural world is governed by divine providence. And we today must deal with the legacy of Kant's "critical" philosophy, according to which everything material is transcendentally ideal and science rather than philosophy is authoritative in all matters empirical. Dealing with this two-fold legacy is my task in this chapter.

## WHERE WE ARE

Primary for us is the world we live in, our world, the world as we commonly regard it. Our commonsense understanding of things is like the air we breathe or the ground on which we walk. Without this milieu we would be at a complete loss. Every inquiry depends on it, either directly, as when you are looking for your keys, or indirectly, as when physicists are conducting an experiment in quantum mechanics.

In the world as we commonly regard it our default position is what Husserl called "the natural attitude,"[1] implicit in which are four assumptions central to our present inquiry. First, that reality does not depend on how we regard it: the assumption of metaphysical realism. Second, that there exists a natural world: the assumption of realism with respect to things natural. Third, that of the many kinds which seem to exist, some are natural: the assumption of realism with respect to natural kinds. And fourth, that our everyday understanding of things is generally adequate: the assumption of pragmatic reliability with respect to common sense. Whereas the first, second, and fourth are very difficult to dislodge by Borgesian, Platonic, or otherwise skeptical provocations, the third is less secure: the phenomena themselves sometimes prompt us to wonder whether seemingly natural kinds are genuine. While trying to answer a pair of questions about such kinds, we tentatively accept the other three assumptions. Tentative acceptance in philosophical reflection is not the same as simple presumption in the natural attitude. To mark this difference, let us call "hypotheses" the following premises on which our inquiry relies:

H1. Reality does not depend on how we think or speak about it.
H2. There exists a natural world.
H3. The way we ordinarily think and speak about it has heuristic value.

H1–H3 belong explicitly to the suppositional nexus in which we now undertake to investigate natural kinds.

For those of us trained in philosophy, however, the situation is complicated by the fact that in this more rarified milieu there are two doctrines standing in our way, each threatening to stop us dead in our tracks. One is a sort of antirealism, prevalent in philosophical circles today, according to which reality depends on how we conceive it. This goes directly against H1 and indirectly against H2; it denies *natural* philosophy its proper subject matter. The other obstacle is a sort of naturalism, also prevalent today, according to which we should take for granted the ontology implicit in scientific theory, since it alone provides reliable guidance in any investigation of nature. This goes directly against H3, and impedes the renewal of natural *philosophy* today. To proceed, therefore, we must first remove these roadblocks, or at least find them to be less formidable than they seem. Let us try cutting them down to size. And while doing so, let us bear in mind the following. First, the antirealism and the naturalism standing in our way have both originated in Kant's critique of theoretical reason: the former, in his transcendental idealism; the latter, in his scientific realism. For Kant, these were but two sides of one doctrine, whereas for us they are decidedly two. Thus has the Kantian crisis evolved, and those trying to revive natural philosophy here and now must confront both.

Second, these objections, like their common ancestor, are highly theoretical, one drawing us into the recesses of metaphysics, the other into the nature of philosophy. Although both go against the grain of common sense, one cannot refute them by appealing to this rougher ground without begging the question in favor of H1–H3. Nor, for similar reasons, can one refute them by noting with horror the moral and political consequences that seem to follow if we accept either. Nor can one refute them by casting suspicion on the fact that they accord wonderfully well with some recent trends: antirealism, with the relativism popularly espoused in certain contexts; naturalism, with the authority popularly ascribed to natural science. The objections now before us are properly philosophical and must be dealt with accordingly.

## ANTIREALISM

Some antirealists deny the reality of this or that (sort of) thing and some deny that reality is independent of how we regard it. The former are metaphysical antirealists, the latter transcendental or protometaphysical antirealists. Call them *m-antirealists* and *pm-antirealists* respectively; and their opposites, *m-realists* and *pm-realists.*[2]

With respect to natural kinds, then, we have four options: m-antirealism with pm-realism, m-realism with pm-realism, m-realism with pm-antirealism, and m-antirealism with pm-antirealism. Heraclitus and Descartes accepted the first; Aristotle and Kripke, the second; Putnam, the third; Kant, the fourth. Each is a package deal whose parts cannot be disassembled and remixed at will. Take, for example, the natural attitude, which harbors the belief that some natural kinds exist (a variety of m-realism) together with the presumption that what exists is independent of how we regard it (pm-realism). To separate the former from the latter and combine it with pm-antirealism would be like putting the "same" smile on a different face or weaving the "same" episode into a different tale: the transformation effected is obvious as soon as we notice the contexts. One cannot be a Putnamesque or Kantian pm-antirealist *and* an Aristotelian or Kripkean natural-kinds realist.

We can better describe the subject matter of protometaphysics by introducing a few more technical terms. In philosophy and science, as in everyday life, one distinguishes between being and seeming, reality and appearance, what is real and how it is construed in some epistemic, semantic, or pragmatic situation. Call the former of each pair *ontic,* the latter of each *epistemic.*[3] And call the distinction between them *the onto-logical difference,* the relation between them *the onto-logical relationship.* Whereas metaphysics is mainly about the ontic, epistemology (semantics, pragmatics) mainly about the epistemic, the subject matter of protometaphysics is the onto-logical situation which metaphysics and epistemology alike presuppose. The debate between pm-realists and pm-antirealists is about how one is to understand the onto-logical relationship. Again there are four options: (1) the ontic is independent of the epistemic and conversely; (2) the ontic is independent of the epistemic but not conversely; (3) the ontic depends on the epistemic and conversely; and (4) the ontic depends on the epistemic but not conversely. Accept (1) or (2) and you are a pm-realist. Accept (3) or (4) and you are a pm-antirealist. Some pm-antirealists reduce the ontic to the epistemic. Examples include Dummett, when he claims that truth is just rational assertability, and Goodman, when he claims that worlds cannot be distinguished from world-versions.[4] For them, the ontic is in the epistemic as a tale is in its being told or as a dance is in the dancing. And some pm-realists reduce the epistemic to the ontic. Examples include Neoplatonists, when they claim to have become one with the One, and Parmenides, when he says that thinking differs not at all from being.[5] Such reductions are too extreme insofar as they erase the onto-logical difference and thus obliterate the subject matter of protometaphysics, which philosophy requires. More plausible are the positions of those pm-antirealists such as Putnam, who refuses to reduce truth to rational acceptability, or Butchvarov, who refrains from implying that existence *is* just material identity[6] ; and of those pm-realists such as Plato, when he has Socrates emphasize the "distance" between knower and known, or Parmenides, when he says that discursive thinking is *about* being.[7] These philosophers all agree that the ontic is epistemic-transcendent, and yet, whereas the realists take it to be epistemic-independent as well, the antirealists do not. It is this disagreement, apparently genuine, which concerns us now.

Pm-realism is implicit in the natural attitude. Scientists accept it. So too did philosophers until Kant argued for pm-antirealism.[8] Today there are many pm-antirealists, and so strong do their arguments seem that pm-realists now find themselves on the defensive. How did this happen? What motivates so many philosophers to reject what is deepest in our everyday sense of reality, in scientific theory, and in much of the philosophical tradition? For some, it has been a way of preserving necessary truths about the world, or any *world* at all, in the withering light of Humean empiricism. For others, growing up in the wake of logical positivism and its demise, it has been a way of taking into account such things as the underdetermination of theory by observation, the inseparability of theory from the circumstances of its discovery or construction, the theory-laden character of empirical evidence, the seeming impossibility of separating the world from what we impute to it in our worldviews. An example from the history of philosophy will help us to see the problem. Aristotle held that horses and the like are individual substances, whereas matter, though involved in all such things, is not itself a substance at all. Descartes held that the substance of things is matter itself, everywhere the same, whereas horses and the like are but transient configurations of matter in motion, not substances at all. So they disagree. By what argument did Descartes establish his own metaphysical position and dismiss the Aristotelian alternative? In *Le Monde* he had recourse to imagination; in later writings he either assumes the correctness of his position or shows how right it seems once you adopt certain epistemological strictures. In his metaphysics of nature, then, the reality of homogeneous matter is less argued for than posited. And how did Aristotle argue for *his* position and against the alternatives known to him? He showed that if in agreement with common opinion you regard horses and the like as individual substances, you cannot then regard their material constituents as modifications of a single substance, matter; but nowhere did he show that you would be right to regard horses and the like as substances. In his metaphysics of nature, too, the reality of such things is less argued for than hypothesized.

It is difficult to decide between Cartesian and Aristotelian conceptions of nature, and the same is true of almost every other major dispute in metaphysics. Theism versus atheism, realist versus antirealist theories of universals, an ontology that admits three-dimensional continuants but not four-dimensional occurrents versus an ontology that maintains the opposite: in no case does there seem to be an argument *pro* or *con* that resolves the issue once and for all. Hence these issues never die, but fade away and bloom anew. Why? The old answer is that reality loves to hide, that the truth in such matters is very difficult to discover, so difficult, in fact, that skepticism remains a serious threat. The new answer is that these metaphysical disputes have about them the air of great depth and difficulty because we presuppose pm-realism. Give up that presupposition and you will see that Descartes did not *discover* something about nature itself that Aristotle missed; instead he brought about a "paradigm shift" by *deciding* that nature was henceforth to be construed in his way, not Aristotle's.[9] This, of course, was not how he understood what he was doing but how we understand it when we adopt pm-antirealism, a doctrine that enables us to diagnose the indecisiveness of the arguments in these metaphysical disputes and free ourselves from the skepticism that has heretofore shadowed them.

Should we then accept pm-antirealism, or stick with the alternative? Again we have arguments *pro* and *con*. Let us briefly consider those advanced by pm-antirealists in recent decades, and see whether they constitute an insurmountable barrier to our nascent inquiry.

### *The Uniqueness Argument*

If reality is fixed, there can be only one description of it that is both true and complete. But there are alternative descriptions wherever we look. In physics one may describe spacetime in terms of different reference frames, its fundamental constituents in terms of particles or fields, and there are diverse interpretations of quantum mechanics. In geometry one may regard points as primitive elements or as limits of convergent sets. And in metaphysics, given objects *a*, *b*, and *c*, one may take them to be three in number (as we usually do) or seven (as mereologists do). Though incompatible, such theories are both acceptable, and their diversity is ineliminable; there can be more than one description of reality that is both true and complete. Therefore reality is not fixed, as pm-realists suppose.[10]

Reply

If saying that reality is fixed is a just way of saying that it depends not at all on how we conceptualize it, then it does not follow that only one description of it can be both true and complete. From the "fixity" of the ontic, little follows regarding the epistemic. In particular, pm-realists need not reject conceptual relativism altogether. They may distinguish entities from objects of interest and argue that many such objects do not exist; or that no metaphysical description can be adequate since reality transcends discursive thought; or that none can be complete since there are non-denumerably many sorts of entity, too many to name in any language having a finite alphabet; or even that some Kantian or non-Kantian categorial scheme is alone adequate. The uniqueness argument fails to admit these and other options because it mistakenly assumes that the fixity of the ontic entails the fixity in principle of the epistemic.

*The Truth Argument*

Pm-realists are committed to a correspondence theory of truth. But if truth consisted in a correspondence between thought-signs and thought-independent things, there would be no way of deciding when this relation obtained and when it did not; only by assuming a "God's-eye view" could one make the needed comparison between thought and thing, and no such view is available. Hence the very idea of truth would have no discernible application. Better then to accept something like a rational-assertability theory of truth.[11]

Reply

The pm-realist is not committed to any specific theory of truth. When asked to say what truth is, we tend to think of it in terms of matching, fitting, or corresponding. Truth is like that, we say. Maybe this analogy is not as good as we think. Then we should look for a better analogy. In any event, pm-realists need not take truth to be a relation between epistemic items "in here" with ontic items "out there." They may develop a proto-epistemology in which thought-signs have no place, and truth itself they may regard as ontic or epistemic or somehow both. In any case, this argument about *truth* obscures the issue at hand, namely, the mind-independence or mind-dependence of *reality*.[12] Like the uniqueness argument, it assumes that one can refute pm-realism by focusing not on the ontic but on something broadly semantic.

*The Inscrutability Argument*

If the ontic were mind-independent, as pm-realists suppose, we could have no epistemic access to it. Nor would there be grounds for believing that it exists; the very idea of mind-independent entities would have no application. We could still think and talk about such things, of course, but it would be like peering into a pitch-black space and describing invisible objects supposedly there; our "descriptions" would be idle speculations, not genuine statements. Pm-realism posits a reality that is too remote ever to come within the purview of our epistemic resources. It mistakes recognition-transcendence for recognition-independence. Pm-antirealism does not make this mistake. It can bring reality closer to us, where it was all along, without reducing recognition-dependence to recognition-immanence.[13]

Reply

Goldbach's conjecture may be undecidable in Peano arithmetic and Cantor's continuum hypothesis is undecidable in standard set theory, yet neither is empty. And if some statement should be absolutely undecidable, what reason might we have for deciding a priori that it is truth-valueless?[14] Be that as it may, our pm-antirealist now assumes that pm-realism must take the ontic and the epistemic to be mutually independent. Some old-time realists do speak as if mind and reality were like two locally separate containers, one filled with thought-sentences, the other with entities—a view which old-time skeptics happily exploit.[15] But they can and should say instead that the epistemic depends somehow on the ontic, though not conversely. Then the inscrutability argument loses much of its apparent strength.

*The Model-Theoretic Argument*

Suppose we have a theory *T* which under ideal epistemic conditions we would accept. Then, having formalized the language of *T*, we find that it has indefinitely many models (interpretations or schemes of reference under which all the observation sentences of *T* are true), only one of which may have been intended at the outset. Which model do we choose? Neither our formalized theory nor a theory-independent world can tell us. Therefore, contrary to what the pm-realist assumes, the predicate signs of *T* have no determinate extensions.[16]

Reply

It is doubtful that anyone's actual understanding of reality can be completely formalized, yet without this step the argument before us does not get off the ground. Yet suppose the formalization requirement can be met, and that the pm-realist should meet it. Suppose also that if the epistemic is to be described in representationalist terms, then pm-realism is untenable. Must we grant the antecedent? No, for there are nonrepresentationalist alternatives.[17] This argument also cuts both ways, for what now is to count as *rational* acceptability, our pm-antirealist's new criterion of truth? It has to be epistemic rather than ontic, as do the activities of formalizing, interpreting, finding that experiential and operational and theoretical constraints are met, and so on.[18] Rationality is involved throughout, and though not without some ontic accompaniment it cannot be fully entified.[19] Pm-antirealists need to clarify their notion of rationality and show how the rational acceptability of one model over another is to be determined. Only when they have done so will we be able to tell whether their position is any more tenable than the one they reject.

*The Identifiability Argument*

We take some objects to exist, others to be non-existent. We also take some apparently distinct objects to be numerically identical, others to be non-identical. In taking two or more objects to be numerically identical, we judge them to be one and the same entity. The concepts of existence and identity are applied as a pair. And since the former is even more obscure than the latter, in protometaphysical reflection we must understand existence in terms of numerical identity, not the reverse. No two objects are themselves numerically identical, since as objects they are two and not one, nor are they numerically diverse as the single entity each of them *is*. The concept of identity is applicable in this situation, but it does not stand for anything ontic or epistemic. Employing it is something we just *do*. The same is true of the concept of existence. It is by employing this pair of concepts that we "move" from the epistemic to the ontic. Apart from this transcendental activity, reality would be nothing to us.[20]

Reply

This is the strongest of the five arguments I have canvassed.[21] Yet it too is not decisive. First, identification may not be the only way across the onto-logical divide; Plato, Aristotle, Descartes, and others speak of an intellectual intuition in which one is directly, non-conceptually "in touch" with the ontic. Pm-realists today seem ready to join their opposites in pooh-poohing any such "God's-eye view." This, however, is but a feature (an unfortunate one, in my opinion) of the current debate, and not essential to pm-realism in general. Second, the "concept" of existence is not obviously paired with the concept of identity in the way this argument requires. Notice that our world includes such things as utensils, laws, persons, countries, and other institutions. Though these putative entities are publicly identifiable, we may have philosophical reasons for judging them to be non-existent. We may even conclude that *our* world is mostly a collective dream, a transcendental production within Plato's Cave. In any event, the link between existence and identity is hardly less obscure than existence itself.[22]

There we have it: five arguments against pm-realism, none conclusive. The first four are mutually supporting, yet their collective strength is not enough to stop us dead in our tracks. Nor is the fifth strong enough to hold us back. We are free then to go on in our investigation of natural kinds, or would be if philosophical naturalism were not still impeding our way. Before turning to this second obstacle, however, a final note about these five arguments. All rest on two key assumptions: first, that the epistemic is independent of the ontic if, as pm-realists claim, the ontic is independent of the epistemic; and second, that the epistemic and its relationship with the ontic must be understood in a way that undermines pm-realism. A few pm-realists accept the first assumption, but they need not and should not do so. Many more accept the second assumption when they buy into the picture of the epistemic that pm-antirealists paint for them, but they should think long and hard before doing so. For what could be more resistant to elucidation than the epistemic and its relationship with the ontic? Has anyone, realist or antirealist, so well understood the protometaphysical situation that she can declare, "This is how it is!"? I think not.[23] Indeed, pm-realism and pm-antirealism appear to be the thesis and antithesis of a meta-Kantian antinomy—the mother of all antinomies—irresolvable by a trans-transcendental turn, perhaps irresolvable altogether. The effort to get beyond intractable metaphysical disputes by recourse to pm-antirealism has landed us in an intractable protometaphysical dispute—which again is progress of a sort—and in the indecision it brings we are free to set aside pm-antirealism for the time being. Eventually we may have to reconsider our decision to go on hypothesizing pm-realism (H1) and robust m-realism with respect to nature (H2). That would not be surprising: philosophy is always circuitous, the ground on which it proceeds always shaky.

## NATURALISM

Our primary resource, the naturalism of common sense, gave us H1 and H2. The naturalism now gaining favor among philosophers accepts these two hypotheses, but on grounds having less to do with common sense than with recent developments in philosophy. First, the linguistic turn no longer grips philosophers as it once did. In ethics they are once more undertaking normative studies; in semantics they have revived Mill's theory of direct reference; and in philosophy of science they no longer feel constrained to examine only the sciences that examine nature. Nature itself, the great outdoors, is once more an object of philosophical study. Metaphysics is back out of the doghouse.[24] Second, the chasm separating philosophy from science has been partially overcome. At any rate, those calling themselves "philosophical naturalists" or "physicalists" have decided to join the scientists by climbing aboard their ship. True, they were not invited, and once made aware of their presence scientists tend to regard them as parasites. Yet quite apart from matters of etiquette, one wonders what role if any these philosophers can play. It remains the job of natural scientists to discover how and why natural things are as they are. Their theories are the best explanations we have. And we, presumably, are to accept them as authoritative until scientists advance something better. So what remains for these philosophers to do? Plenty, it seems. On the one hand there is *ontological* naturalism: one assumes the ontology of physical theory to be approximately correct and explores such topics as the nature of spacetime, the necessitarian character of physical laws, the relation between such laws and causal powers or natural kinds, or between entities posited in different natural sciences.[25] On the other hand there is *epistemological* naturalism: one assumes that something exists just in case it is natural and proceeds to "naturalize" things seemingly nonnatural by showing that physics requires no reference to numbers or universals or God, or by reducing epistemology to a chapter in empirical psychology.[26] In short, having boarded the ship of science and subordinated themselves to the resident theorists in charge, these naturalists proceed along philosophical paths marked out by and for themselves.[27]

It is in this way, above all, that natural philosophy being renewed today. Its practitioners accept pm-realism (H1) and m-realism with respect to nature (H2). So far, then, they agree not only with the scientific attitude they have adopted as philosophers but with the "natural" attitude I regard as my primary resource. But they reject H3, the hypothesis that folk naturalism can be very helpful in trying to understand things natural, whereas I now accept it; and some of them accept the proposition that anything extra-natural is nothing at all, whereas I do not. Here then is a second major obstacle to renewing natural philosophy in the way I propose. Is it, too, less formidable than it seems?

Consider first the proposition that whatever is not natural is nothing at all. Construed protometaphysically, it is unacceptable. For again, the epistemic (semantic, pragmatic) cannot be absorbed into the ontic without eliminating the onto-logical distinction and with it the possibility of philosophy or science. To suppose that epistemology can be fully naturalized is to blind oneself to the protometaphysical situation without which the naturalization of epistemology could not occur. Granted, much of what we call "knowledge" is explainable in terms of natural states and the flow of information. But science will never fully account for the first-person character of our primary claims to know or of our foremost practical and theoretical inquiries. The reason is simple: Scientific theory is rigorously third-person or impersonal. It has no room for "I" or "we," "here" and "now."[28] These terms occur only informally in the patter with which a scientist presents her theory and reveals its context of discovery or construction. Such terms prove indispensable, moreover, when trying to describe the protometaphysical milieu in which the real appears. Philosophical naturalists should think harder about the difference between "I think" and "He thinks," "here and now" and "spatiotemporal location *p*," "my body" and "body *a*"; between the protometaphysical statement, "We are humans," and the metaphysical statement, "Individuals *a* and *b* are humans." By so doing they will see that the onto-logical cannot be reduced to the ontic, which in turn will enable them to make better sense of their own reductionist activities and become more astute in their disputes with pm-antirealists.[29]

Construed metaphysically, the proposition in question means that something exists only if it is natural. Is this true? Though scientists may use it as a heuristic maxim, it cannot be deduced from any of their theories, not indeed from the so-called theory of everything. Scientists themselves prefer to make the more modest claim that some putative entity (say luminiferous ether) is not required to explain natural phenomena (the wavelike character of light).[30] That something exists only if it is natural is a philosophical claim whose truth value is debatable. In chapter 3 I will argue that universals exist, that they are nonspatiotemporal entities, and that any such entity is transnatural. This too is debatable. At the outset, then, we should not assume that such things exist or agree with philosophical naturalists when they insist *a priori* that something exists only if it is natural. The naturalism I propose is at once more liberal and more circumspect than theirs.

This brings me to their rejection of H3. By simply taking for granted the ontology implicit in scientific theory, these naturalists have made a methodological error. For in philosophy, ontological understanding comes at the end of inquiry, if at all, not at the beginning. And it has to be *our* inquiry, not another's. True, we must get our bearings in order to begin, but we do so by accepting certain beliefs as working hypotheses or clues, not as articles of faith. It is also true that we today would be foolish not to include among such clues what scientists have to say about the natural world, since they are in some sense authoritative on the subject; but what they say is for us authoritative hearsay, and even the most authoritative hearsay is not good enough when one wants to *know*. In natural philosophy, then, one should not simply assume that the naturalism of science is even approximately correct.

The philosophical naturalist may object as follows. "While for us, at least, no scientific theory is cognitively adequate, such theories are far and away our most valuable resource when investigating things natural. We should then replace my H3 with H3′: With respect to the natural world, our only reliable source of hypotheses is the disciplinary matrix of natural science, since it alone puts one in touch with the natural world as it is."

To answer this objection, let us first review the relation between science and philosophy. Only recently has the difference between them become obvious. Descartes, Huygens, and Newton were all natural philosophers. And although he distinguished between physics and metaphysics, Leibniz did not think that his *Specimen dynamicum* was a work in anything other than philosophy. And although Locke and Hume were doing other things, neither of them regarded Newton as anything other than a philosopher. Only later in the eighteenth century did philosophy and science separate. Only then did natural philosophy seem to have no subject matter of its own.[31] Philosophers henceforth focused on questions in epistemology, logic, ethics, politics, and theology, and began reaching out to encompass things cultural and historical. By the late nineteenth century, the line dividing philosophy from science had become a chasm. We today are caught up in this history. Anyone who wishes to renew natural philosophy in a traditional sense of the term, as I do, must deal with the problem it poses.

So how should we deal with it? Consider the following analogy. By the mid-nineteenth century, philosophy and science had diverged to the point they were like two distinct species. Subsequent variations within each have made the differences between them still more pronounced. Yet signs of their common ancestry persist. On both sides one finds a strong interest in discovering the truth; an emphasis on the need for disinterested inquiry, for evidence, and on the role of argument; a strong tendency to dismiss out of hand all appeals to authority, and to disregard preestablished boundaries on what it may and may not investigate. The differences are important, but so too the generic identity. Now as Darwin observed, it is between closely allied species that the competition is fiercest. He was thinking about survival, of course, whereas we are thinking about the possibility of natural philosophy. Philosophical naturalists recommend one sort of rapprochement: philosophy is to become scientific, or at any rate parasitic on science in a way that poses no threat to the host. I recommend, instead, that we renew natural philosophy by continuing to do our own thing, disregarding the commonly held opinion that philosophy has no business investigating the natural world, and using whatever seems to us to be the case about nature, and whatever experts tell us about nature, as clues in our attempt to understand it. This is what philosophers have always done, regardless of what the resident authorities have instructed them to believe.[32]

Why then do we make the folk naturalism of common sense our primary resource? There are two interrelated reasons. First, as noted in chapter 1, the concept of a natural kind has its native home in common sense rather than scientific theory. Indeed, Kripke and Putnam almost grant as much when they extend their new theory of reference to "natural kind terms"; for as examples they mentioned "water," "gold," and "tiger," commonplace words which have acquired a theoretical use, and not such words as "quark," "photon," and "Hox C-8 gene," which originated in scientific discourse and have not yet acquired a colloquial use.[33] These natural-kinds realists are, as we shall see in chapter 6, pre-scientific essentialists who take their referential bearings by ordinary discourse. Second, recall that our questions about natural kinds arose in wondering about the world as we commonly understand it. Consequently, in trying to answer these questions, we are less likely to commit fallacies of irrelevance if we stay with common sense for a while, before turning to relevant scientific theories and philosophical traditions. Eventually we may find that our pre-theoretical understanding of things has less heuristic value than we had hoped, but we have no way of knowing this at the outset.

In sum, our nascent inquiry depends on several hypotheses. Protometaphysical antirealism and philosophical naturalism threatened to nip it in the bud, the first by refuting H1 and removing the robust sense of reality in H2, the second by dismissing H3 as unreliable. Having now diminished if not dismantled these post-Kantian roadblocks, we are free to proceed in under the gentle guidance of common sense. Our turn to nature has begun.[34]

## NOTES

1. Husserl 1962, §§ 27–30; 53f, 64f.
2. I borrow the term "protometaphysical" from Butchvarov 1979. My use of it differs little from his. My use of the term "metaphysical" also differs from his, and from Putnam's. What I call "protometaphysical realism," Putnam calls "metaphysical realism" and what he sometimes calls "internal realism," I call "m-realism within a pm-antirealist framework." Most philosophers are both local m-realists and local m-antirealists, affirming the existence of some putative things, denying the existence of others.
3. What I call epistemic you may prefer to call proto-epistemic. In any case, do not take the (proto)epistemic as such to be a second reality alongside the first, or to be unreal in the sense that unicorns and illusions are unreal. Appearances as such are not realities, nor are they necessarily "mere" appearances.
4. M. Dummett, e.g., "Truth," "Realism," and "Bringing about the Past," all in Dummett 1978; and Goodman 1978, 4, 94, 96, et passim.
5. Parmenides B3. That *noein* here is non-egoic shows just how far Parmenides' understanding is from that of Descartes or Kant.
6. Putnam 1981, 55; "Vagueness and Alternative Logic," in Putnam 1983, 275; and his 1989, 115. Consider also Butchvarov 1979, chapter 2.
7. Cf., e.g., Plato, *Phaedrus* 247c3ff, and Parmenides, B 8.51.
8. Perhaps one should count Protagoras as a predecessor.
9. Cf. Kuhn 1970, 103, 151, 158, 199–204; and his 1977, 139, 173, 204–07, 322–39.
10. Putnam 1981, chapter 3; "Equivalence," in his 1983; 1987, 18–22, 32f; 1990, 98; and his 1995, 109–23. Also Goodman 1978, chpt. 1. Analogous arguments have been made concerning the interpretation of a text.
11. Putnam 1981, 55f, 122f, 216; and "Introduction," "Reference and Truth," and "Convention: A Theme in Philosophy," all in his 1983, viii-xii, 81ff, 177. A similar argument can be made regarding reference. My rejoinder would be similar to the one made here.
12. M. Devitt has rightly emphasized this in his 1991, passim.

13. H. Putnam, e.g., "Why There Is No Ready-Made World," in Putnam 1983, 207, 210. Also Dummett 1978, xvii–xx, xxiii, xxx, 8, 21, 233, 364, and Wright 2000, 361.

14. Khlentzos 2004, 98–104, 147f.

15. Consider analogous disputes about the inscrutability of a transcendent being. For some today, the natural world seems as remote cognitively as God did for some medieval theologians, partly because they accept the causal theory of perception. Neither realists nor antirealists need accept this theory; and even if they do, pm-antirealism does not immediately follow: Kitcher 2001, 156ff.

16. Putnam, "Models and Reality" in 1983; also his 1981, 217f with 33–35, "Realism and Reason," in Putnam 1978, 123–38; and his 1987, 18. On the relation between Putnam's m-realism and Kant's, cf. C. Brown, "Internal Realism: Transcendental Idealism," in French 1988, 145f, 151, 154.

17. S. J. Wagner, "Why Realism Can't Be Naturalized," in Wagner 1993; and Khlentzos 2004, 225–46, 284–86.

18. The signs in the formalized language of *T* are probably ontic.

19. H. Putnam, "Why Reason Can't Be Naturalized," in Putnam 1983. Reason is trans-historical, he says here (234f), but elsewhere he says that our notion of rationality has evolved (e.g., 1983, vii–viii). So is it both transcendental and immanent, but in different senses?

20. Butchvarov 1979, especially chapter 2, and his 1998, 127, 134, 140, 148, 153f, 164n52. For Butchvarov, as for Wittgenstein, what we just do need not be whimsical or willful.

21. It appears also to make sense of the debate between Aristotle and Descartes, introduced earlier in this section. Aristotle, in the interest of contemplation, said, "Let animals and other material things have eidetic natures," and so it was until Descartes, in the interest of mastering nature, said, "Let the corporeal world be mathematizable matter in motion according to laws." Because neither realized what he was doing, each believed he had discovered the way things are. Cf. Nietzsche, *Beyond Good and Evil*, § 211 with § 150; also *Gay Science* §§ 110, 121 and *Will to Power* §§ 481, 521. Also W. James, *Principles of Psychology*, (New York: Dover, 1950), I:271–90.

22. For a more detailed reply, see: S. Umphrey, "Rethinking the Subject Matter of Protometaphysics," in Blackman 2005.

23. One reason for the difficulty seems to be that our cognitive abilities are geared to the ontic, which explains our tendency to reify the epistemic by making it a second reality alongside the first. Perhaps the epistemic can only be shown in one's discourse about the ontic.

24. The expression "great outdoors" I take from Q. Meillasseux, *After Finitude: An Essay on the Necessity of Contingency*, tr. R. Brassier (London: Continuum, 2008), 7 et al. For a fuller account of the naturalistic turn in contemporary philosophy, cf. P. Kitcher 1992, 53–114.

25. Representative works include Armstrong's 1993 and 1997.

26. Representative works include H. Field's *Science without Numbers: A Defense of Nominalism* (Princeton, NJ: Princeton University Press, 1980) and Quine's "Epistemology Naturalized," in Quine 1969.

27. A clear statement of this position can be found in P. Maddy, *Second Philosophy: A Naturalistic Method* (New York: Oxford University Press, 2009).

28. E.g., J. Lear, "The Disappearing 'We,'" *Proceedings of the Aristotelian Society* S48 (1984): 219–42; B. Huebner & D. Dennett, "Banishing 'I' and 'We' from Accounts of Metacognition," *Behavioral and Brain Sciences* 32 (2009): 148f; and, e.g., Kornblith 2013, 122–41. That the third-person is really impersonal: É. Benveniste, *Problèmes de linguistique générale* (Paris: Gallimard, 1966), 225–36, 251–66. Kornblith's account of third-person psychological studies is predominantly first-person.

29. Cf. P. Butchvarov, "Comments," in Blackman 2005, 76f, 96f, 132–35, 156f, 303–07. Also Putnam 1981, 113; S. J. Wagner, "Why Realism Can't Be Naturalized," in Wagner 1993, 211–53; M. Sacks, *Objectivity and Insight* (New York: Oxford University

Press, 2003), 172–77; T. Nagel, *The View from Nowhere* (New York: Oxford University Press, 1986), 54–66; and, e.g., L. R. Baker 2013, 28ff.

30. As for the claim that whatever exists is material, which some physicalists still make, quantum field theorists would dismiss it as laughably out-of-date.

31. Nineteenth-century "natural philosophers" were scientists: Faraday (who published regularly in *The Philosophical Magazine*), Darwin, and Maxwell, among others.

32. Cf. Bealer 1998, 201–39; Foley 1998, 241–56. Philosophers are like those who produce post-colonial literature by "writing back against" the formerly hegemonic and still influential culture from and by means of which they become free. Philosophy is a kind of post-presumptive reflection in which one "thinks back against" the prevailing conceptions of things.

33. Cf. Donellan 1983, 84–104, That they did not also mention as examples of natural kind terms such ordinary words as "air," "mud," or "ghost" is but one of many signs that they take natural kinds to be "Peircean," explicable scientifically rather than commonsensically. This is what Putnam means by "division of linguistic labor."

34. In terms favored by Leo Strauss, we have ascended from the sub-cave of contemporary theory to the cave of the natural world given pre-theoretically. In political philosophy, he claimed, it is *only* by means of historical studies that we can return to political affairs as presented in the twilight that is proper to them. Whatever one may think of this claim, I do not find his historical turn to be necessary or sufficient in natural philosophy.

# THREE

# Universals and Particulars

The concept of a natural kind has its proper employment in typological thinking about things which have come to be and which persist (remain the same) in spite of or because of the many changes they undergo, as long as they exist. In this chapter I first consider the metaphysical basis of typological thinking generally, and then determine the place of natural kinds within this ballpark.

## KINDS AS TYPES

Some kinds are artificial or fictional (tables, teapots, Dementors, Furies), and some are political or conventional (monarchies, oligarchies, Saudis, Columbians). On reflection these all seem to depend on our beliefs and practices. Other kinds of kinds do not. Of these, some are mathematical—the division of conic sections into hyperbolas, parabolas, ellipses, and circles, or the division of "natural" numbers into odd and even, prime and composite—while others seem to exist in the world around us. There are flammable things, water-soluble things, four-legged things and six-legged things, trees and shrubs, metallic things, evergreens, wolves, oaks, salmon, gulls, water, and so forth. It is kinds of this sort, all apparently real, which interest us here.

The word "kind" is now commonly used to designate both an extension or *class* and the character or *type* which determines that class. So, for example, humankind seems not only to be (the class of) all and only humans, but also that which determines the class by characterizing all and only its members. The class-character is the type. Each and every member of the class is a token of that type. The difference between type and token is like the difference between the letter A and any occurrence of that letter in this sentence. In each of these cases, humankind and A-kind, members may come and go and yet the kind itself remains the same. In neither case does the type determine how many are in the class, nor does the identity of the class depend entirely on the identities of its members. For natural kinds, as for alphabetic kinds, the class depends on there being a type of the appropriate sort, and perhaps at least one token of that type.[1] As types, then, natural kinds exist only if some types are real in the sense that they exist independently of our beliefs and practices.

Real types and their tokens are called "universals" and "particulars." Universals are multiply exemplifiable entities; particulars, the things which exemplify them. To clarify what this means, consider several flowering plants. Let them be two bearded iris and one sunflower. And to avoid this tendentious way of singling them out, label them $\alpha$, $\beta$, and $\gamma$. We notice that $\alpha$ resembles $\beta$ much more closely than it resembles $\gamma$. We classify them accordingly. But to which similarities and dissimilarities do we attend when making this division? It is hard to say. Indeed so complicated are $\alpha$, $\beta$, and $\gamma$ that philosophers today prefer to simplify their paradigmatic particulars. Following suit, let us attend not to these three plants as organic wholes but to their floral parts *a*, *b*, and *c*. Then, focusing on the shape and hue of each, we see that *a* is iris-shaped and violet whereas *b* is iris-shaped and yellow, and *c* is sunflower-shaped and yellow. Such is our paradigmatic situation. There are three particulars—*a*, *b*, and *c*—or six if you count the shapes and hues proper to each—S*a* and H*a*, S*b* and H*b*, S*c* and H*c*. Do *a* and *b*, though numerically distinct, exemplify one and the same shape? Do H*b* and H*c* instantiate one and the same hue? If the answer to either question is "yes," then, in addition to the two mentioned particulars, there is the universal which each exemplifies or instantiates. But if the answer to such questions is always "no," then there are no universals, no real types.

One may worry that our paradigmatic situation is still too complicated.[2] But the more serious issue is this: We say that *a* and *b* have their shape in common or share the same shape, or that the color of *b* recurs in *c*. And in describing our paradigmatic situation we use general terms such as "shape," "hue," "iris," "sunflower," "floral part," and "plant." Our very discourse seems to favor realism with respect to universals. As noted in chapter 1, however, our everyday understanding of things does not commit us to this metaphysical thesis, nor to its antithesis. Moreover, as a moment's reflection will show, neither thesis nor antithesis can be deduced from the phenomena themselves. Perceptual situations of this sort prompt us to ask whether universals exist; they do not answer our question. To find the answer we must go beyond phenomenology to ontology. We must subject the folk metaphysics of common sense to philosophical reflection.

## THEORIES OF UNIVERSALS

On the issue before us philosophers have articulated the full range of options. There are two main camps. On the antirealist side we have unqualified nominalism and resemblance theories. On the realist side we have one-in-many theories, according to which universals are immanent in particulars, and one-over-many theories according to which they transcend particulars. Debates among these four positions are ongoing. My aim here is quite modest: to determine whether it would be reasonable to accept tentatively, at least, some version of realism; for only then might it also be reasonable to affirm the reality of natural kinds as types and not merely as classes.[3]

### *Unqualified Nominalism*

What I call "multiple exemplifiability of a universal" is nothing more than the multiple applicability of a predicate or concept to particulars. Thus in our paradigmatic situation *a* and *b* belong to the class of iris-shaped things, *b* and *c* to the class of yellow things, simply in virtue of the fact that "[is] iris-shaped" applies to *a* and *b* but not to *c*, while "[is] yellow" applies to *b* and *c* but not to *a*. The division of things into particulars and their properties, and their allocation to different sorts, are linguistic artifacts produced by an operation analogous to application. We classify particulars simply by labeling them and not by comparing them with a blueprint, paradigm, or template. To think otherwise is to be bewitched by language or gripped by conceptual illusion.

Difficulties besetting unqualified nominalists include the following. (1) They cannot explain coextensive kinds. Take for instance "equilateral triangularity" and "equiangular triangularity." The corresponding predicates apply to the same objects, yet their meanings are neither the same nor logically equivalent. How is this not a difference between coextensive properties?[4] (2) They cannot explain the applicability of predicates to new cases. Suppose we know how to use the word "yellow" by having learned its application to *b* and *c*. Then, on observing the sun, we know that "yellow" applies to it as well, since otherwise we do have not yet learned how to use the word. What accounts for this novel application if not the fact that the sun is yellow? How is it that we know how to go on? (3) What is the status of *linguistic* types? Take for instance the word "predicate," which seems to recur in nominalistic discourse. Are these tokens of a single type, or do they only seem to be such because a second-order predicate (concept) applies to them? Here nominalists face a dilemma: either they admit real linguistic (intra-mental) kinds or they do not. If they do, their position appears incoherent. And if they do not, they must accept an endless series of higher-level predicates which, though not viciously regressive, has no basis in our actual language or conceptual scheme. (4) They must clarify the falling-under relation between particulars and applicable predicates. If *a* and *b* fall under "[is] iris-shaped," while *b* and *c* fall under "[is] yellow," then these two relations fall under "falling under," as does this third relation, and so on to infinity. Do they accept this infinite series, or do they reject relations as well, including the falling-under relation? (5) They are unfaithful to the phenomena. For similarities and dissimilarities between *a* and *b*, *b* and *c*, are evident to sight. There is evidently some basis in reality for our classifying *a* and *b* as F-shaped things, *b* and *c* as G-colored things.

*Resemblance Theories*

Kinds are nothing more than classes of mutually resembling particulars. These particulars are qualified things such as *a*, *b*, *c*, or else properties such as S*a*, S*b*, S*c*, H*a*, H*b*, H*c*. The latter are called "tropes" and the resemblance theorists who affirm their existence are called "trope theorists."[5] Those resemblance theorists who deny the existence of tropes grant that *a*, *b*, and *c* are qualified in many respects, but deny that these qualities are entities in their own right. All agree that the only entities are particulars and that they resemble one another, more or less, quite apart from how we think or talk about them. They should also agree that a resemblance between two entities is not a third entity over and above its relata, that exact resemblance (if there be such) cannot be reduced to numerical identity, that the resemblances among existing particulars are exceedingly complex, and that degrees of resemblance are difficult to quantify even on an ordinal scale. To deal with this difficulty, some resemblance theorists require that class members resemble one another to some minimum degree. Others introduce exemplars to serve as clear cases of class membership. For instance, let *c* and the sun be exemplars for the class of yellow things. Then, since *b* (H*b*) resembles *c* (H*c*) at least as much as *c* and the sun resemble each other, *b* too belongs to the class of yellow things, whereas *a* (H*a*) does not. We may even regard *a* as a Quinean "foil," a clear case of nonmembership in the class.[6] To discern and distinguish these resemblance classes we appeal to predicates or concepts, some of which we may form on the spot, but never to kinds as real types or universals, since there are no such entities.

Difficulties besetting resemblance theories include the following. (1) In the absence of any real types, our choice of exemplars appears somewhat arbitrary. So too, in the absence of any resemblance test, do resolutions of our disagreements about whether *x* resembles *y* at least as much as *y* resembles *z*.[7] There are also borderline cases whose classification would be arbitrary. Resemblance theorists may grant all this; but to the extent they do, one wonders how much basis in reality there is for the tidying-up we do while classifying particulars. Might it not be better to describe their many similarities and dissimilarities and leave it at that?

(2) As Bertrand Russell and others have noted, if *x* resembles *y* and *z* resembles *w*, these two relations resemble each other, this third relation resembles each of the other two, and so on to infinity.[8] And as Michael Loux and others have noted, this infinites series, though not viciously regressive, has no evident basis in the phenomena.[9] (3) One may reject this argument on the ground that resemblance is an internal relation. But internal relations, too, must have definite numbers of relata. Suppose *x* resembles *y*. It is, we say, a two-term relation. But as Panayot Butchvarov has pointed out, this resemblance statement is necessarily comparative and therefore essentially incomplete; it makes sense to say that *x* resembles *y* only if one presupposes some *z* and *w* such that *x* resembles *y* more than *z* resembles *w*. This statement is complete but it is not a resemblance statement. Hence, what we call "resemblance" is no relation at all.[10] (4) Resemblance theorists are empiricists. Even if there were universals, they say, we could never be aware of them. Not everyone agrees. Kurt Gödel, for one, admits the possibility of perception-like intuitions of mathematical objects.[11] And among these objects, apparently, are universals. Do we not sometimes intuit such entities in the course of mathematical study? I think so. No resemblance theorist has argued successfully to the contrary.

### *One-in-Many Theories*

Universals exist in the particulars which exemplify them and thus resemble one another. For instance, in *a* and *b* there exists the universal $U_S$ in virtue of which they are rightly said to have the same shape. In other words, the numerically distinct quality-particulars S*a* and S*b* instantiate one and the same $U_S$. But one-in-many theorists understand these quality-particulars in two very different ways. Some construe them protometaphysically: the relationship between S*a* and S*b*, or between each of them and $U_S$, is like the relationship between the Evening Star and the Morning Star, or between each of these celestial objects and the planet Venus. As distinct particulars, then, S*a* and S*b* are not real, while as one and the same entity each of them *is* $U_S$.[12] Others, including Gustav Bergmann, construe them metaphysically: the so-called quality-particulars $S_A$ and $S_B$ are really the distinct states of affairs S*a*′ and S*b*′, each analyzable into the universal $U_S$, a particular which may but need not be *a* or *b*, and a nexus of exemplification binding these two constituents into a single state of affairs. The relationship between S*a*′ and S*b*′ resembles that of two intersecting circles which have a part in common and two parts in virtue of which they differ.[13] Yet, however they understand quality-particulars (this-suches), these theorists are all immanent realists; universals exist *in rebus*, in particular things and not apart from them. Must they agree that universals are spatiotemporal entities? No, for it may be that their (nonreal) instances are spatiotemporal whereas the universals they instantiate are not, or that states of affairs have their feet planted firmly in the spatiotemporal world while their heads are in a nonspatiotemporal domain. Must they agree that a universal exists only if actually exemplified? So it appears. Some may also require that it be exemplified at least twice, whereas others may require only that it be exemplified somewhere somewhen.

Difficulties besetting this position include the following. (1) Perceived particulars do not wear universals on their sleeves: $U_S$ has no shape at all, and $U_C$ is arguably colorless. Are we then nonperceptually aware of the universal that a quality-instance *is* or that a state of affairs contains?[14] If not, if protometaphysically or metaphysically immanent universals are *cognitively* inaccessible, on what basis does one accept a one-in-many theory? (2) How exactly are we to understand the relationship between particulars and the universals "in" them? Regarded protometaphysically as the onto-logical relation "between," say, *Sa* and $U_S$, it shares in the obscurity of onto-logy generally. Regarded metaphysically as the real nexus of exemplification uniting $U_S$ with *a′* in *Sa′* it appears subject to the Bradley Regress: there must be something uniting this nexus with that universal and with this particular, and so on to infinity. Though not viciously regressive, this consequence is nonetheless unacceptable.[15] (3) Exact resemblances are regarded as identities. How then are inexact resemblances to be regarded? Perhaps *a* and *b* exemplify the same *specific* shape-universal; or if, as is likely, their shapes are only roughly similar, perhaps they exemplify the same *generic* shape-universal. As we shall see presently, however, species are real types only if genera are not. (4) Universals are multiply exemplifiable entities. Whence the necessity that they be exemplified? It cannot be in the universals themselves or in the particulars which exemplify them. It must then be in the (proto) metaphysical situation involving both. But how?

### *One-over-Many Theories*

Universals exist beyond the particulars that exemplify them and thus resemble one another. So, for instance, in addition to *Sa* and *Sb* or *Sa′* and *Sb′* there is a transcendent universal $U_S$ in which these particulars participate and are therefore rightly said to be two phenotokens of the same onto-type. Because universals are transcendent rather than immanent, one-over-many theorists cannot take the relation between *Sa* or *Sa′* and $U_S$ to resemble that between the Morning or Evening Star and Venus, or between whole and part. It more nearly resembles the relation between images and their original, or between aspiring individuals and a goal they can never attain. One-over-many theorists easily agree that universals are nonspatiotemporal. They may but need not require that these multiply exemplifiable entities be exemplified.

Difficulties besetting this position include the following. (1) The relation between a phenotoken and its onto-type is obscure. It is no more obscure than the nexus of exemplification (posited by some immanent realists), or the relation between a quality-instance and what it *is* (posited by other immanent realists), or the *sui generis* relation of resemblance (posited by qualified nominalists), or the relation between a word and that to which it applies (posited by unqualified nominalists). But that it has problematic company does not make it any less problematic. (2) Transcendent universals, whether exemplified or not, seem cognitively inaccessible. Notice, however, that metaphysical transcendence does not entail epistemological transcendence. The moon is no less evident in its watery reflections than Venus is in its twilight presentations or the property of being S is in *Sa′*, *Sb′*, etc. Notice also that, just as Hume was able to infer a missing shade of blue, so too we may be able to infer the existence of never-exemplified universals. In Euclidean geometry there are cubes; in hyperbolic geometry there are horospheres; and in elliptical geometry there cannot be figures of either sort. Since the geometry of the physical world is elliptical, there are no cubes or horospheres in it, yet we can recognize cubicality and horosphericity by studying alternative geometries. Let us not, however, underestimate the epistemological difficulties attending a one-over-many theories, particularly if it allows the participation of phenotokens in their onto-types to occur in degrees, as representations of the moon are more or less distorted because of surface turbulence, the funhouse mirror in which they appear, or the seismic nature of our minds. To appreciate this difficulty, imagine before you an ordered array of figures varying gradually in shape between a rectangle and an ellipse. Let the variations between neighboring members is small but perceptible. Now select a figure from among those in the middle and ask, "Is it a borderline phenotoken of rectangularity, or of ellipticality, or does it exemplify some other shape-universal?" No fact of the matter, it seems, would enable you to answer this question. For us, at least, transcendent universals may not be reliable principles of classification: there is too great a slip between the cup and the lip.

Having now surveyed the positions for and against the reality of universals, let us try to decide which if any is acceptable.[16]

## A FOURTH HYPOTHESIS

There appear to be objective similarities and dissimilarities among existing things. Unqualified nominalists try to explain away this appearance. That this proves very difficult to do may be why most nominalists are actually resemblance theorists. Let us agree that there are objective similarities and dissimilarities. Our task, then, is to explain them metaphysically. Resemblance theorists try to explain them without admitting universals, whereas immanent (*in rebus*) theorists and transcendent (*ante res*) theorists admit universals but disagree on the question whether such entities must be instantiated or exemplified. How is one to weigh the difficulties attendant on each of these three theories, so as determine which of them bears the least burden of proof? That this proves very difficult to do may well be why the debates *pro* and *con* remain indecisive. Again we find ourselves desiring to know without yet knowing; and in this state we are free to accept, if we wish, any one of them. I tentatively accept realism, making it our fourth explicit hypothesis:

> H4. There are universals, i.e., nonspatiotemporal multiply-exemplifiable entities which are somehow in or beyond the spatiotemporal things which exemplify them.

H4 fully accords with H1–H3. Notice, moreover, that only if this hypothesis is true can there be natural kinds as real types and not merely as classes. Indeed, as I shall argue in chapter 5, there can be natural kinds in the full sense of term only if the one-over-many (*ante res*) theory of universals is true. But we are not yet ready for that argument, so I omit for now any further restriction of H4. Already, however, we can see that this hypothesis, though necessary for the reality of natural kinds as types, is hardly sufficient since it tells us nothing about which putative universals are genuine. How might one begin to fill in this blank?

## EIDETIC SPACE

Universals appear to be many in number and to constitute a sort of system. Call the nonspatiotemporal place of this system "eidetic space." We realists have the task of identifying its contents and describing the relations among them, of mapping eidetic space. We may thus discover whether some real types are natural kinds.

The usual way of identifying universals is linguistic: from the existence of predicates such as "___is red" or "___is a color," and of abstract nouns such as "triangularity" or "animality," one infers the existence of corresponding universals. It has been argued that our employment of such items *cannot* be explained unless universals exist. Even granting as much, however, one need not and should not conclude that every abstract noun names a universal; for if it did, every principle of classification would turn out to be a real type, including those whose role is evidently practical rather than theoretical and those we may invent as we please. So, if we are going to explore eidetic space linguistically by way of meanings, we must find some nonlinguistic way of anchoring our conjectures in reality.

Philosophers have recommended approximately three ways of keeping it real: by appealing to intellectual or eidetic intuition, or to broadly empirical constraints, or to broadly logical or conceptual constraints. The first way was endorsed by Plato and Aristotle. Philosophers today generally dismiss it, but it merits reconsideration. The second way is favored by naturalistic and scientific realists who, taking things spatiotemporal to be instantiations of universals, use them as so many clues to things genuinely eidetic. D. M. Armstrong is a strong proponent of this approach.[17] So was Aristotle. And so too am I, though unlike Armstrong and others, I see no need to turn immediately to natural science, and like most natural-kinds realists today I see no need to return to Aristotle.

The third way is mainly negative: if some putative universal is found to be logically or conceptually incoherent, it does not exist. Realists have used this test to rule out squircularity (being at once both round and square) and the like. They have also used it to rule out negative universals (notcircularity, masslessness, etc.) and disjunctive universals (being round-or-square, grueness, etc.). Of course, we have no reason to expect that conceptual space, particularly as we conceive it, will turn out to be isomorphic with eidetic space. It turns out, moreover, that conceptual incoherence is sometimes hard to determine. Consider the following.

Are there higher-order universals? The answer would seem to be "yes" if there are adjectival, relational, and substantival universals; and in any case, all universals seem to exemplify universality. Is this higher-order universal self-exemplifying? If not, there would appear to be an infinite regress of higher-order universals. But since that consequence is unacceptable, let universality be self-exemplifying. There are then, two classes of universals: those which exemplify themselves (e.g., universality) and those which do not (e.g., circularity). Take the class of nonself-exemplifying universals and ask whether nonself-exemplifying universality belongs to it. If it does it does not, and if it does not it does. This is a version of Russell's Paradox. Perhaps it can be thwarted here, but the issue remains murky.[18]

Are there conjunctive universals? The answer appears to be "no," since any such entity would have two or more *entities* as constituents and thus be *analyzably* complex, and one cannot posit such complexity in a universal without entering the Bradley Regress. One might think that coordinate determinates (specific universals) could be conjunctive in the sense that each shares a determinable (generic universal) with the other—in somewhat the way a right-angled triangle and a rectangle may so overlap that they have a trapezoidal figure in common.[19] But as I have argued elsewhere in detail, if the determinates are two entities, the determinable they are said to have in common cannot be a third entity; and if the determinable is an entity, the determinates said to specify it cannot be additional entities.[20] Eidetic realists, therefore, have a choice. Pluralists tend to accept species and reject the generic universals said to be in them, whereas holists tend to accept "genera" and reject the universals said to specify them. Some may even conclude that so-called universals, though seemingly many in number, are really so interrelated that they constitute a single complex universal which the world as a whole exemplifies.[21] In any event, which choice we *should* make remains unclear.

Finally, are all metaphysical necessities to be explained eidetically? Suppose one thing is crimson, another pink. The former is darker than the latter, necessarily. Is this necessity grounded simply in the color-universals so exemplified, or is it rather the case that these universals are types of hue, that the instantiations of such types must come with certain degrees of brightness, and that this *non*-eidetic condition of hue-instantiation explains the fact that tokens of crimson are necessarily darker than tokens of pink. Or suppose something has a circular surface. Every circle must have a circumference and a diameter. Furthermore, the ratio of its circumference to its diameter must vary not at all with size and be equal to the number $\pi$, or else vary with the size and have $\pi$ as its lower limit, or else vary with the size and have $\pi$ as its upper limit. These differences follow not from circularity but from the nature of the space in which circles exist.[22] Should we say that in addition to circularity there are types of spatiality, and that exemplification of the former requires exemplification of some one of the latter? Or should we say that in addition to circularity there are non-eidetic conditions of its exemplification? The answer remains unclear.

Natural kinds appear fraught with such difficulties. First, supposing some natural-kind types to be genuine universals, is there a second-order universal (namely, natural-kindhood) which they and they alone exemplify? If so, is it yet another natural-kind type? Second, should we be pluralists or holists with respect to natural kinds themselves? Third, if there are non-eidetic as well as eidetic conditions of exemplification, are not tokens of natural-kind types likely to be exemplary manifestations of them? Let us, however, put aside such difficulties until we have considered more closely the particulars which theories of universals generally presuppose.

## THEORIES OF PARTICULARS

With respect to particulars, too, antirealism is an option. One may take the instances of universals to be nothing but spatiotemporal appearances—(proto)epistemic rather than ontic, and as such neither real nor unreal—and reduce concrete things to complexes of such instances. Alternatively, one may take every concrete thing to be nothing more than the set of compresent universals it exemplifies, and reduce exemplification to the higher-order relation of compresence which accounts for the unity of the set. A difficulty attending the former alternative is that these instance-complexes do not seem so firmly rooted in their spatiotemporal locations that they could never be identified and reified over time.[23] Among difficulties attending the latter alternative are these: the identity of indiscernibles is not preserved; the notion of compresence is obscure; and a unified set of universals, even one that includes types of spatiotemporal relatedness, is neither spatiotemporal nor particular in the way that red spots or tigers are spatiotemporal and particular.[24]

Suppose there are real particulars. They come bare or clothed, and if clothed, thin or thick. Let us briefly describe and appraise these three options.

### *Bare Particulars*

According to Gustav Bergmann and others, if two numerically distinct things exemplify one and the same quality-universal, there must be in each a particular that is itself unqualified, without any nature whatever, and that differs simply from every other bare particular. Entities of this sort are at once the ultimate exemplifiers of universals and the ultimate relata of spatiotemporal relations. There is no reason to think any of them persists.[25]

Difficulties besetting this theory include the following. First, it is hard to make sense of the exemplification relation in virtue of which particulars appear to be qualified. Second, aside from the fact that they become dialectically indispensable when one decides to undertake a Bergmannian assay of things perceived, bare particulars are cognitively inaccessible; we have no awareness of their existence. Though one may argue that acquaintance with bare particulars is somehow possible, the difficulty here should not be underestimated. Realists with respect to universals and particulars generally begin with what they perceive, which they take to be at once qualities-of-particulars and instances-of-universals. The task, as they see it, is to distinguish universals from their instances and particulars from their qualities. Ensuing difficulties come in pairs: (1) to make sense of the instantiation relation and the inherence relation, (2) to distinguish universals from their instances and particulars from their qualities without putting them beyond our ken, and (3) bring them together without turning universals or particulars into non-entities. To avoid the second pair of difficulties, realists today generally reject bare particulars as well as transcendent universals. The question is whether they can affirm immanent universals or essentially qualified particulars without having to downgrade their status as entities.

### *Thin Particulars*

According to trope theorists, what I call "quality-instances" are entities in their own right. Every such trope is both qualified and particular, a this-such, but without being analyzable into a quality and that which has it. Hence no tie is needed to hold a trope together: the Bradley Regress is precluded. Tropes, moreover, are thin in the sense that each is one-sided, nonsubstantial, and not composed of more elementary tropes. There is no reason to think any of them persists.[26] According to D. M. Armstrong, what I call "quality-instances" are really states of affairs. In every elementary state of affairs there is a particular and the universal it exemplifies; and this particular is thin in the sense that it and the universal it exemplifies are internally related, partially identical constituents of the single state of affairs to which they belong. Hence no tie is needed to bind them together: the Bradley Regress is again precluded. Being essentially clothed, moreover, particulars are evident in the states of affairs to which they belong: the principle of acquaintance is not violated. And by belonging to the states of affairs into which the spatiotemporal world divides, universals do not exist is a realm apart: scientific realism is preserved. Finally, although its constituents are internally related, a state of affairs may nevertheless be contingent. There is no reason to think that the particular in it persists.[27]

We thus have two diverse ways of understanding thin particulars. Both are problematic. For trope theorists, a major difficulty concerns the fact that tropes may coincide. Take for instance H*a*, S*a*, and S*b*. Whereas H*a* and S*b* are locally separate (in virtue of relational tropes, which are additional independent entities?), H*a* and S*a* are not (in virtue of still other relational tropes?). Is there any fact of the matter enabling one to tell how many tropes, even how many "H*x*"s, exist right there right now? Apparently not. For Armstrong, a major difficulty concerns the status of universals. If the basic entities are states of affairs and not the particulars or universals in them, and if every such entity is unanalyzably complex in the sense that its constituents are internally interrelated, then universals are also too "thin" to be entities in their own right.[28] To preserve his realism regarding universals, he must deny that they depend internally on this or that instantiating particular; but then their "immanence" is incidental to the universals themselves. Armstrong, then, must either admit that universals as such are but possible state-of-affairs types or else give up his naturalistic thesis that something exists only if it is spatiotemporal. We see by this turn of events how ontology can be like trying to straighten out a warped board: you fasten it down here and it pops up again over there.

### *Thick Particulars*

In everyday life we encounter individuals and masses that seem irreducibly thick in the sense that each is many-sided and substantial, capable of changing and being changed in its surroundings, and of persisting even though it changes qualitatively, quantitatively, and locally—wherefore there appears to be in it some basis for distinguishing between substantive and attributive, the what and the how. Such things are for us exemplary entities. But *are* they irreducibly thick? Many philosophers think not. Some reduce them to bundles of tropes, others to collections of Armstrongian or Bergmannian facts, and no such multitude of nonsubstantial entities is a single concrete individual or stuff. It seems, however, that dogs, water, and the like cannot be explained away in this manner; nor is it clear that they are the complex substances they appear to be. Realism with regard to such things remains an open question.

Suppose there exist some irreducibly thick particulars. How is one to understand them? Three options are available to us. One, introduced by Locke and revived recently by C. B. Martin, posits an enduring substratum in which properties inhere. These attributes (the nonrelational as well as the relational) are all incidental; they come and go, yet the thing continues to exist as long as its substratum persists.[29] The major difficulty confronting this option, of course, is that the substratum itself is as bare as a Bergmannian particular and thus quite beyond our ken. Another option, suggested by Husserl and elaborated by Peter Simons, takes concrete individuals to be trope-like complexes whose constituents are not mutually independent tropes but rather dependent "moments" of a single entity. The Bradley Regress is thus precluded. And to account for the fact that this entity persists even as some of its attributes are replaced, one may posit in it an enduring nucleus, a trope-like complex within, and maintain that the other moments depend essentially on it but not conversely.[30] Simons, however, has only proposed this as a metaphysical possibility. It remains to be seen whether any such particulars exist. And even if they do, they seem too thin, too insubstantial to be dogs, walleye, or the like.

The third way, introduced by Aristotle and revived by Michael Loux and E. J. Lowe, posits enduring substances to which attributes belong but regards any substance itself not as a bare substratum or a bare this, but as fundamentally a this-such, a determinate nature or essence; and whereas the incidental and necessary attributes of particular substances instantiate *attributive* universals, their essences instantiate *substantive* universals. There is then some basis in reality for the categorial distinction between adjectival expressions ("iris-shaped," "canine," etc.) and sortal terms ("iris," "dog," etc.). Whereas tokens of attributive universals could not exist apart from the individuals to which they belong, tokens of substantive universals may exist apart from some but not all of their incidental attributes. And again, these substantial tokens are just the individuals themselves, *what* they are essentially, whereas the others constitute *how* these individuals are, necessarily as well as accidentally.[31]

M. J. Loux and E. J. Lowe seem to have given us what we seek: there are irreducibly thick particulars of the sort presupposed in everyday life; some, if not all, of these particulars belong to natural kinds in virtue of their essences; and these kinds, as types, are real substantive universals. Both of our questions seem to be answered. But not so fast. First of all, the existence of individual substances is presupposed, even though realism with regard to such things has not yet been established. That's one difficulty. Another is that no account has yet been given of masses such as water and gold, which, considered as scattered wholes or as isolated portions, do not appear to be tokens of sortal universals.[32] A quick way out of this difficulty would be to postulate two sorts of substantive universal: those exemplified by individuals and those instantiated by masses. But perhaps the concept of a natural kind is properly sortal, and applies to masses only by analogical extension. Or perhaps apparent masses are really collections of individuals. Or perhaps apparent individuals are really portions of masses. Is it not conceivable that the exemplifier we seek is substantial spacetime or an ever-changing field of energy whose distinguishable regions we mistake for real particulars? If monistic mass-ontology seems incredible to us today, that is probably because a preference for object-ontology has become habitual even among philosophers.[33]

A third difficulty is that from the evident distinction between sortal and nonsortal terms, or the inevident distinction between essence and attribute, it does not follow that there are two types of universal, namely, substantive and attributive. In geometry we take triangular areas to be triangles; in physics, whatever has a small set of nonrelational properties to be an electron; in chemistry, whatever has the structural property $H_2O$ to be water; and in biology, as in daily life, we take canines to be dogs. Laymen and experts alike tend also to reify tokens of salient functional types: refrigerating contraptions are refrigerators, encoding regions of RNA structured thus and so are codons, some indeterminate loci of electromagnetic force are electrons, etc. The grammatical distinction between attributive types such as *triangularity* and *caninity* and substantive types such as *trianglehood* and *doghood* is certainly worth noting. It may or may not have some basis in reality. Suppose it does. Then that basis may be eidetic—in which case these two types of universal are themselves real second-order universals—or non-eidetic, having to do with the attribute-essence distinction that is necessary for something to be a particular substance—in which case these two types are not real universals at all. Which is true here remains to be shown.[34]

Finally, this way of understanding thick particulars is for us a *doctrina ex machina*. As with the nomological account of natural kinds considered and set aside in chapter 1, so too with the neo-Aristotelian alternative now before us: We have not yet done the hard work of making it ours, and we wish not to repeat the error of those natural philosophers who, having found light to be wavelike, set about describing the special medium that *had* to exist if light-waves were to be propagated. Their description of this ethereal medium was constrained by the role it must play in explaining the phenomena, by the limits of broadly logical possibility, by the need for coherence with the rest of physics, and little else. Similarly, advocates of the option before us, having decided that individual substances exist and that some of them belong to natural kinds, set about describing the special universals they must exemplify and the underlying essences which instantiate them. Their account, too, is constrained by the role these postulated entities must play, by the limits of broadly logical possibility, by the need for coherence with the rest of their ontology, and little else. In both cases, the explanations (one physical, the other metaphysical) are initially very attractive. But there are also gaps and loose ends. We wonder whether in this case, too, the whole thing will collapse like a house of cards.

There are then at least four reasons for not accepting the neo-Aristotelian option as it stands. Irreducibly thick particulars remain something of a mystery.

## AGAIN FROM THE BEGINNING

To see how we might go on, let us review. Natural kinds are conceivable as classes and as types. As types they are real just in case they are universals, i.e., multiply exemplifiable entities. To clarify what it is to be a universal and find out whether universals exist, we began with three specimens of flowering plant—two bearded iris α and β, and a sunflower γ—yet so complicated and confusing was this situation that we focused right away on their flowers *a*, *b*, and *c*, or on the shapes and colors of these parts, S*a* and S*b*, H*b* and H*c*, etc. Thus were we able to distinguish between realism and antirealism with regard to universals, to distinguish varieties of each, and to find that realism of some sort is acceptable. We made it our fourth hypothesis and noted that, while it does not entail the reality of natural kinds as types, there are natural kinds in the full sense of term only if H4 is true. Then, having surveyed the domain of universals, we returned to the particulars that exemplify or instantiate them. Again we distinguished varieties of realism and antirealism, especially with regard to substantial individuals. This part of our inquiry was less successful: we could not decide whether such particulars are irreducibly thick or not, and if not, whether they should be regarded trope-clusters (as a thing-ontologist would) or as states of affairs (as a fact-ontologist would). Regarding what seem to be natural-kind tokens—sunflowers and dogs, water and gold—our inquiry so far has led us into an impasse.

Looking back, however, we see that almost from the beginning our inquiry has been broadly logical. Proceeding *logikôs*, as Aristotle put it, befits the issue of universals. Aristotle himself proceeded in this way in his *Categories* and in his *Metaphysics* VII.4ff. The Scholastics followed suit. So too have realists and antirealists in our time. But the issue of universals is not the same as the issue of natural kinds. To clarify what it is to be a kind of this sort, and find out whether there exist any such kinds, we must also proceed naturalistically, or *phusikôs*, as Aristotle put it. He himself did so in his *Physics* II and assorted biological writings. We should do likewise by undertaking an investigation that is complementary to the one just outlined; complementary in the sense that it is naturalistic rather than broadly logical and bottom-up (from things to their kinds) rather than top-down (from real types generally to their tokens). In other words, having exploited modern theories of universals and particulars—proposed by philosophers most of whom have not been *natural* philosophers—we should now reorient our inquiry along lines suggested by D. M. Armstrong and a few other philosophical naturalists.

Unlike Armstrong and others, however, I will not turn immediately to the world of natural science, but instead continue proceeding in accordance with hypothesis H3 (that our everyday understanding of the world is heuristically valuable). Consequently, for the time being at least, I will rely on the folk metaphysics of common sense, our primary resource, whose usefulness we have not yet fully exploited. My lingering with prescientific opinion may seem odd to those philosophical naturalists who, notwithstanding my argument in chapter 2, still have trouble accepting H3 even as a hypothesis. I ask them please to be patient and see for themselves what the present inquiry yields.

In everyday life, such things as flowering plants $\alpha$, $\beta$, and $\gamma$, birds and fish, water and the like are certainly real individuals. We found this presumption to be problematic, but our loss of innocence does not prevent us from taking seriously the world as ordinarily understood. According to our everyday understanding of things, individual animals and plants are among the paradigmatic entities. They also seem to be members of natural kinds. So let us begin again with them; let us go directly into the thicket of things, armed only with common sense and our ability to think.

Making full use of our primary resource requires not only close observation of things paradigmatically real for us but also attention to ways in which we ordinarily understand what is observed, which are reflected in the semantic fields of the words "nature," "natural," and "kind." The more we investigate them, the more entangled these fields appear. "Natural," for example, is cognate with "nascent" and "innate." It is no great surprise, then, to find that in earlier English the word "nature" was also used to designate generative fluid and the female genitals. And having noticed the relation between "kind" and "kin," and found that they are cognate with "genus" and "generation," it is no great surprise to find that in earlier English the word "kind" often meant nature (as in "ones kinde" and "God of kinde") and was used to designate families, ethnic groups, and generative fluid.[35] Though some of these uses have fallen away, so that the expression "natural kind" no longer appears pleonastic, the semantic field of this expression still bears traces of its entwined ancestry. It reminds us that natural kinds are *natural* not insofar as they are types, but insofar as the members of such kinds are by nature involved in generation and change, in *genesis*.[36] Bearing this in mind will help us to resist the temptation, ever present in natural philosophy, either to split being (the invariant) apart from becoming (variation) or else to subsume one in the other. Thus forewarned, I will proceed in chapter 4 to investigate things that are paradigmatic entities for us, every one of which seems to change and yet remain the same as long as it exists. Then, in chapter 5, I will try to answer the two questions that have guided us almost from the beginning: What is it to be a natural kind, and are there any natural kinds? Thus will I reconnect what philosophers today have tended to take apart and treat separately: the metaphysics of natural kinds (as types or classes) and the metaphysics of continuants (as irreducibly thick particulars or substances).

## NOTES

1. In mathematics, something exists if it is constructible or otherwise determinable. Hence the set of prime numbers, for instance, is not the sort of class that could gain or lose members. It is possible, then, not only to define the class in terms of the members it has, but also to define the type *primality* in terms of all and only the tokens constituting this class. This extensional turn now prevails in mathematics. Leaving aside the question whether it was wise to make this turn, it appears that natural kinds as classes cannot be defined extensionally—which is yet another reason for wondering whether they can have a secure place in mathematical physics.

2. Given their different blooming times, one can hardly present the floral parts of bearded iris and sunflowers together for visual inspection. Further, the shapes of the two iris flowers do not appear to be exactly alike, nor do the hues of the yellow iris and sunflower petals appear uniform. One may also note that the sunflower's inflorescence is not a single flower, and that I have not distinguished carefully between genus and species, and that hues as we experience them may not be real at all.

3. More detailed arguments for and against universals can be found in Butchvarov 1966; Loux 1970, 3–185, and his 1978; Landesman 1971; Armstrong 1978 and his 1989; Moreland 1985 and his 2001; Fales 1990, 147–253; Swoyer, 2005.

4. Sober 1982, 183–89. In absolute geometry (Euclidean but without the fifth postulate), one *proves* that all equilateral triangles are equiangular. Of course, one may so modify the system that the converse is proved instead. Either way, one must rely on postulates as well as definitions and principles of formal logic, and postulates are not logical truths, since their denials are not formally contradictory.

5. Cf. Williams 1953, 3–18, 171–72; Campbell 1990, 56ff.; Bacon 1995.

6. Price 1953, chpt. 1; Quine 1969, 119f.

7. By "resemblance test" I mean a procedure analogous to the Moh's scratch test for hardness. Without some such test, one does not have even an ordinal scale for ascertaining degrees of resemblance, and without some such scale one has no principled way of forming disjoint coordinate classes.

8. Husserl 1970, I: 410–14; and Russell 1975, 96f.

9. Loux 1978, 45–47, Butchvarov 1979, 196–99, and Moreland 2001, 38–40.

10. Butchvarov 1966, 101–34. Compare, e.g., A. Denkel, "Real Resemblances," *Philosophical Quarterly* 39 (1989): 36–56.

11. K. Gödel, "What is Cantor's Continuum Problem?" in *Philosophy of Mathematics*, ed. P. Benacerraf and H. Putnam (New York: Cambridge University Press, 1983), 383f. See also R. Tieszen, *Mathematical Intuition* (New York: Springer, 1989).

12. E.g., Butchvarov 1979, 184–211, and R. Grossmann, *The Existence of the World: An Introduction to Ontology* (London: Routledge, 1992), 30–45.

13. E.g., Bergmann 1967, 3–21, R. Grossmann, *Categorial Structure of the World* (Indianapolis: Indiana University Press, 1983), 102–54, and J. P. Moreland 2001, 97–100. Few, if any, trope theorists are one-in-many theorists as well; but if they are, and if they regard tropes as entities, they must accept some theory of universals.

14. Cf. Husserl 1962, §§ 3f, 69f, 137, 149; Fales 1996, 99f.

15. Butchvarov 1974, 3–37.

16. The following account supposes that pm-realism is true, and consequently that with respect to universals either m-realism or m-antirealism is true. Compare, e.g., Strawson 1979, 10.

17. Armstrong 1978b. Armstrong's realism becomes "scientific" only after he has established that (immanent) realism is correct, and after he has rejected negative and disjunctive universals and admitted relational or structural universals. In the beginning, then, his account is an exericise in "first philosophy" rather than philosophical naturalism.

18. Cf. Bealer 1982, 94–97. The issue remains unclear not only because Bealer's argument is inconclusive but because the suspected incoherence may lie elsewhere. Notice, for instance, that we now taking first-order universals to be nonspatiotemporal *objects*. Frege and Aristotle have warned us against making this supposition.

19. Armstrong 1978b, 111–15, 117–24.

20. Cf. Umphrey 2002, 134–37.

21. Cf. Fales 1990, 219; Ellis 2001, 249–53.

22. These three options are Euclidean, Lobachevskian (hyperbolic), and Riemannian (elliptical), respectively.

23. Cf. Butchvarov 1979, 212ff, and 1998, 89ff.

24. Cf. Russell 1940, chapter 6; and, e.g., Armstrong 1978a, 91–101. Russell did not assert that particulars are therefore unreal; that is my inference.

25. G. Bergmann, e.g., "Russell on Particulars," in Bergmann 1954, 197–214; and his 1967, 22–25, 112, 117.

26. Most trope theorists are resemblance theorists as well. Yet trope-types (universals) and tropes (trope-tokens) may both exist, and a few trope theorists admit both.

27. Armstrong 1978a, 109, 113–27; his 1997, 8, 28–31, 39–41, 84–88; and his 1998, 103–15. The transition from Bergmannian to Armstrongian facts is a good example of Hegelian dialectic at work in analytic philosophy: The "real" components of the former have been "sublated" into the "ideal moments" of the latter. What were two mutually independent entities, formerly bound together into a derivative fact, have become a single basic fact whose constituents are dependent sub-entities.

28. Cf. Moreland 2001, 83–96.

29. Locke 1979, II.23.4; and Martin 1980, 3–10.

30. Husserl 1970, III; and Simons 1994, 558–75, esp. 567–69.

31. Loux 1974, 773–84; his 1976, 53–61; and his 1978, 159, 163. Also Lowe 2006, 100; and Ellis 2001, 92, 97f.

32. Cf. Strawson 1959, 202, 217–21.

33. H. Laycock is a notable exception—cf. his 1972 and 2006—but his interest is semantic rather than directly metaphysical. Notice that process-nouns generally resemble mass-nouns in not having a genuine plural. A monistic process-ontology, with attendant antirealism regarding individuals, would today be as unusual as a monistic mass-ontology with attendant antirealism regarding particular objects.

34. Consider Wittgenstein 1968, §§ 371, 373; Quine 1960, 210; Armstrong 1997, 34ff; and see also chapter 1 of this volume. Advocates of the position I am now examing regard "tadpole" and "woman," as sortal-nonsortal hybrids. That seems reasonable enough, yet someone with an understanding deeper than ours may regard "frog" and "human being" as likewise hybrid, and so on indefinitely.

35. OED, s.v. "nature," "natural," "kin," "kind." In the Middle English of Chaucer's *Canterbury Tales* these uses are all on display. Kind as nature: KnT I.2, 245, SqT V, 469, 608, PriorT VII.2, 650, CanYPr VIII.2, 659, Melibee VII.4, 1567, et al. Unkind as unnatural: MLT II.88. Kind as family: 2NunT, VIII.1, 121, and probably ParsT X.2, 279, 884.

Kind as generative fluid: ParsT X.2, 965. Kind as sort: KnT I.2, 1401, CanYT VIII.2, 981, ParsT X.2, 448. Nature as sort: ParsT X.2, 678. Nature as generative fluid: ParsT X.2, 577. When distinguishing sorts of morally good or bad dispositions, the Parson uses "specie" (species) rather than "kinde": X.2, 83, 101, 407, 448, 731, et al. "Kind" as "nature" can also be found in early New English; cf., e.g., Shakespeare, *The Merchant of Venice*, 5.1.386, *All's Well That Ends Well*, 1.3.10, *Antony and Cleopatra*, 5.2.262. See also C. Lewis and C. Short, *A Latin Dictionary*, s.v. *natura*, 2D. Notice, moreover, that our words "kind" and "kin" are cognate with words meaning "family," "race," or "tribe," and that from some of these cognate forms come our words "generation," "genesis," "genus," "genealogy," and "general."

36. Compare the very different connotations of "type" and "token," which philosophers today (following C. S. Peirce) use in their accounts of universals and particulars, and which I have used in this chapter. Most modern accounts of universals and particulars are not about natural kinds as such, whereas most premodern accounts of them are about natural kinds most of all.

# FOUR

# Continuants

Entities that remain the same while changing are called "continuants." The concept of a continuant is much contested among philosophers, some even arguing that no such entity *could* exist. In everyday life, however, we take continuants to be exemplary realities. And among those we take to be exemplary as continuants are trees and shrubs, insects and vertebrates, and stuff such as wax, water, or gold. Let our exemplary continuant be a living thing of the last-mentioned sort. Our first task is the get acquainted with this vertebrate—to engage, in other words, in a bit of natural history. Then, having done so, our task will be first to determine what a continuant is, and then to decide whether in fact there are any entities of this sort. What follows is an exercise first in natural history, then in metaphysics, and then in epistemology. All three exercises belong to natural philosophy as I conceive it.

## MEET BLACKIE

Several summers ago a black squirrel showed up in our backyard. We called him "Blackie." Blackie was distinguishable from his grey brethren in other respects as well. For instance, unlike them, he did not try getting at our new "squirrel-proof" birdfeeder from above, but instead went to the ground directly below and ate whatever seeds the birds dropped. He also made daily trips up and down a nearby oak tree and criss-crossed the lawn, pausing occasionally to eat or look around. In general, he stayed away from the other squirrels, who seemed to regard him as an interloper, and succeeded in making our backyard part of his environment for an entire summer.

Though sufficiently distinctive that we were able to recognize him with ease, Blackie was in most respects very similar to other squirrels.

His tail was bushy, his eyes black. When climbing a tree, he would spiral to the side opposite you. When spotting a cat that had entered the yard he would pause on a branch, flick his tail, and make the familiar "chuck chuck kwaach" sound. While coming back down the tree, his hind feet were turned outward in that distinctively squirrelly way, and on nearing the base he would pause to look about, his body still hugging the trunk, his head almost parallel to the ground. Often he hopped across the lawn, his tail making a single arc convex to the ground. While eating something he found, he sat on his haunches, his tail curled up against his back. You get the picture.

Blackie also shared some characteristics so familiar that we easily overlooked them. For instance, though dependent on his environment, he was also independent, doing his own thing in responding to and modifying the world around him. And though changing all the time in location, posture, and size, he remained basically unchanged throughout the summer we observed him. And though partite and complex, behaviorally as well as qualitatively and anatomically, he was no mere collection of bodily parts or characteristics; he had a nature, an inner nature evidently inevident in his parts, features, and way of life.

My summary description of Blackie belongs to the natural history of a continuant—what distinguished it from the others, what it had in common with individuals we call "squirrels," and what it had in common with continuants more generally. Naturalists are interested in characteristics of all three sorts. Many, however, are little interested in characteristics of the third sort, whereas a few are very interested in them. The former are naturalists in the narrower, more usual sense of the term: squirrel watchers, bird watchers, ant watchers, as well as hunters, farmers, and the like. The latter are natural philosophers. It is easy enough to tell the difference between them in the field; for while squirrel watchers are busy trying to find out where animals like Blackie sleep (I was unable to locate it), to determine their sex (that Blackie was male is a conjecture on my part), whether they left behind any offspring, and so on, natural philosophers are wondering what it is to be an individual and whether Blackie really is the individual he appears to be. The difference between these two groups, like that between naturalists and theoretical scientists, is often striking. Yet they are all naturalists. Their differences conceal their commonality, as did the differences between Blackie and the living things around him.

Natural philosophy originates in natural history. It requires comparatively little observation, however, and once underway it may leave its origin far behind. How much natural history do natural philosophers need in order to get started and go on doing their thing? That depends on what they are investigating. For those interested in continuants and their kinds, I suspect the answer is, "Much less than fellow nature watchers tend to believe and much more than philosophers today generally believe." Inquiries of the sort we are conducting require familiarity with the natural world, some of which comes with the folk naturalism of common sense, the rest through direct communion. Ours is an urbanized, highly technological society in which people rarely get outdoors to observe what is there. So paltry indeed is our background in natural history that we are shocked to learn that "primitive" peoples know much more about the natural world than we do.[1] To remedy this deficit, natural philosophers need to remind themselves that communing with nature, like foreplay, is never just foreplay; they need to rejoin the nature watchers with whom they have more in common than either side is wont to believe.[2]

In any case, our paradigmatic situations now are of the sort represented by Blackie in his real-world environment (and mine). We wonder about the fact that this living thing is independent as well as dependent, unchanging as well as changing, and that it has an inner nature as well as outward features. In our wonderment about such things we move beyond natural history to ontology, thus parting company once more with our more observationally-minded brethren, in order determine what a continuant is and discover, if we can, whether Blackie and the like are in fact entities of this sort.

## METAPHYSICAL DIFFICULTIES

A continuant changes and yet remains unchanged. This is contradictory, Aristotle observed, only if the thing it is changing and not changing "at the same time and in the same relation."[3] Blackie both climbs and does not climb a tree, but not at the same time. He is both varying and invariant, but not in the same respect. This was Aristotle's way of allowing for natural entities such as Blackie. The temporal qualification is acceptable unless time is unreal, in which case motion too is unreal; and if motion is unreal, realism with respect to nature (H2) is false. The relational qualification is also acceptable, unless the requisite respects do not exist independently of our distinguishing them, in which case realism with respect to continuants is false. What might these so-called respects be?

The variations in question must be *real* and they must be variations *in* the continuant *x*. Call such changes "intrinsic to x." Blackie's increasing a little in size was a change intrinsic to him. That he first looked small beside grey squirrel$_1$, and then large beside grey squirrel$_2$, was a change extrinsic to him, a mere "Cambridge change."[4] One might think that all relational variations are extrinsic, but this is not so. When Blackie was first hopping or climbing and then sitting on his haunches, these variations in posture were intrinsic to him. It may also be that the ways in which he was related to things around him were not simply extrinsic. Physics teaches us that some intrinsic properties are relational as well: Blackie's weight depends partly on the ambient gravitational field, while his mass and shape depend partly on which inertial frame is assumed to be at rest.[5] The distinction between monadic and relational properties is unclear, and so too is the distinction between intrinsic and extrinsic changes.

Suppose the variations in question are intrinsic to *x*. Then *x* is an enduring continuant only if it remains unchanged; and this invariance, this numerical identity over time, must also be real and intrinsic to *x*, for if it were extrinsic—a mere "Hume identity"[6]—*x* would not be a genuine continuant, since in becoming other it did not endure but ceased to be. To allow for the reality of continuants we must distinguish intrinsic changes that are substantial from those that are not. Continuants remain invariant in and through their nonsubstantial variations, whereas substantial variations mark real discontinuities in which they perish or come into being. There can be no genuine continuants if, as Quine and others maintain, the distinction between substantial and nonsubstantial change depends on the way we conceptualize the world.[7]

The concept of a continuant is coherent only if the distinction between substantial invariance and nonsubstantial variation can be made in a way such that one can hold them together in thought. That this is not easy I now show.

To distinguish substantial from nonsubstantial, some philosophers proceed mereologically, some predicatively. A mereological analysis seems appropriate, given the fact that Blackie has bodily parts, some of which appear more important to his ongoing existence than others. He could survive the loss of a hair, an eye, or perhaps the mereological changes involved in becoming grey; but not the loss of his heart, all or most of his mitochondria, or the mereological changes involved in becoming a blob-like mess on the street. It is difficult however to identify any material part of Blackie that must remain invariant if he is to continue existing. Nor can any part of him explain why he is an individual whole and not a dividual heap; to suppose otherwise leads one straight into the Bradley Regress. Nor, moreover, it is easy to see why those parts that are necessary to him must exist together with those parts that are not. If one separates the two sorts, then Blackie is really a hard-core entity, invariant as long as it exists, plus an ever-changing heap-like cloud of contingent parts, neither of which is itself a continuant; and if one refuses to separate them by requiring that the invariant core be attended by *some* transient parts, then the metaphysical basis of this requirement must be elucidated. Nor, finally, is it easy to see how this mereological analysis could be extended to (putative) masses such as water, a continuant whose portions are not proper parts.

Viewed predicatively, a continuant *x* is distinguishable into its *how* and its *what*. One specifies how *x* is in terms of its attributes. These attributes may or may not be instances of universals; they are in any case ways in which *x* exists—real, inherent properties of the thing itself.[8] And what the thing itself is, one identifies with its substance. The substance of $x$ may or may not instantiate a universal; it is in any case what Joseph Almog has called "a condition of existence," a reality apart from which the attributes of $x$ and $x$ itself would not exist.[9]

Nonsubstantial changes are variations in the attributes of $x$, through which $x$ endures, whereas any variation in the substance of $x$ is tantamount to its demise. The case is complicated, however, by the fact that continuants have necessary as well as contingent attributes. Variations in Blackie's size, posture, or health are nonsubstantial as a rule, but any variation in the endoskeletal or bilaterally symmetrical character of his body-plan would be substantial.[10] Thus, more fundamental than the distinction between how and what, attribute and substance, is the distinction between contingent ways in which a continuant exists (all of which are attributes) and necessary conditions of its existence (which are either attributes or the substance itself). But the distinction between how and what does not thereby vanish, for attributes inhere in the substance $x$, whereas the substance itself does not inhere in anything; it is categorially basic, whereas its properties are not. It would be a category mistake, then, to suppose that the substance itself could be reduced to a set of necessary properties. For this reason, among others, one should not identify the substance called "Blackie" with the alleged property of *being Blackie* or with (the necessarily unique instance of) Blackiehood.

This way of viewing continuant things can explain metaphysically their individuality as well as their dividuality, and how it is that such things both change and remain unchanging as long as they exist. It can also be extended to continuant masses, whose manner of being partite and unitary is quite different. But there are still difficulties. First, it is not easy to distinguish clearly between necessary properties and the substance in which they inhere. Second, the requirement that a continuant have *some* contingent properties must have some basis in the entity itself, yet what that basis might be is hard to discern. Third, it may or may not be that one can identify and track an individual continuant only by employing a (substantive) sortal concept. In any case, the concrete substance $x$ itself is a this-such that cannot be reduced to a bare particular (haecceity) or a common nature (sortal or nonsortal universal) or to a combination of the two.[11] Thus $x$ itself is unique without being so private as to be altogether unlike other substances, and comparable without being so public as to not be unique. This is hard to understand.

Consider once more our paradigmatic continuant. On glimpsing it from a distance you might point and ask, "What is that?" My answer would be, "Blackie," using the name I gave to this particular squirrel. Suppose you then ask, "And what *is* Blackie?" You want know what this now-identified entity is *in its very nature*, its *essence*. "I don't know," I say, yet agree that your follow-up question is a good one, for Blackie not only has a nature, there is a sense in which he *is* that nature.[12] More generally, given a continuant *x* with its inner nature $n_x$, there is a sense in which *x is* $n_x$. Statements of this form I call "nature statements." Examples include, "Water is $H_2O$," "Blackie is genome$_B$," "Socrates is his soul"—even, "The man is his clothes," if only as a degenerate case. Perhaps none of these is true, yet they remind us of similar statements such as, "This stuff is water," "That animal digging up the lawn is Blackie," "Hesperus or Phosphorus is Venus," and "The locally separate color-instances you see before you are the universal *chartreuse*." These I call "protometaphysical essence statements."[13] They too may not be true. Certainly they are mysterious, though no more so than their metaphysical analogues. Let us compare them, for sometimes in philosophy two obscurities are better than one.

Though often linked with identity statements, protometaphysical and metaphysical essence statements are not statements of identity. We do not say that Venus *is* Hesperus or that $H_2O$ *is* water. Nor would we say that Blackie and his intrinsic nature are identical in the way that Blackie-in-the-tree (Arborus) and Blackie-on-the-ground (Terras) are identical. It is true, however, that continuants *x* and *y* are numerically identical if and only if $n_x$ and $n_y$ are one and the same nature or essence. Similarly, Arborus and Terras are numerically identical if each of them *is* the entity Blackie. Exploiting a notational device invented by David Wiggins,[14] we rewrite these two statements as follows:

$$\underset{B}{A=T} \qquad\qquad \underset{n}{x=y}$$

Notice, however, that whereas the former takes us from two perceptual objects to the entity that each of them is, the latter takes us from a continuant entity in different states at different times to the essence that it most of all is. Whereas the former pertains to the *onto-logical* or *protometaphysical* difference and relation between given appearances and some reality, the latter pertains to the *ontic* or *metaphysical* difference between an entity and its intrinsic nature. Notice, moreover, that the difference between a continuant *x* and its nature $n_x$ is such that the former cannot be reduced to the latter, and the relation between them is such that, necessarily, they come into being and perish together. Notice, finally, that $n_x$ explains the unity-*cum*-invariance of *x* in *x*'s multiplicity-*cum*-variations.[15] It follows that $n_x$ should not be regarded as a continuant, for then it too would have an intrinsic nature and so on to infinity.[16] Nor should every nature statement be regarded as a protometaphysical essence statement, for without the metaphysical difference between *x* and $n_x$ there could be no continuants and the concept of a continuant could have no real application. Our task is to elucidate this difference without turning it into something it is not.[17]

## METAPHYSICAL DIFFICULTIES (CONT.)

In its approach, if not in every detail, the foregoing account is rather traditional, going all the way back to Aristotle. I now propose a rather novel way of understanding continuants, one which begins from the fact that they are all involved, to some degree, in the natural world to which they belong.

Things natural are changing all the time; they share in the commotion of the world. Call such involvement *metabolic*. They are also partite, sharing in the materiality of the world. Call such involvement *compositional*. And they depend for their existence on things around them, remote as well as as proximate, past as well as present. Call such involvement *situational*. Universals are not involved in the spatiotemporal domain of their exemplifications either metabolically or compositionally or situationally. They are not natural things. Must everything natural be involved in all three ways? The universe itself is not situationally involved. Electrons do not appear to be compositionally involved. And so brief are the "lives" of some other subatomic particles that their metabolic involvement seems to consist in nothing more than coming into being and directly passing away. It is doubtful, however, that any of these counterexamples is a genuine continuant. Exemplary individuals such as Blackie and exemplary masses such as water exist metabolically and compositionally and situationally. They are, in virtue of their three-fold involvement in the natural world, categorially other than the trans-natural universals they exemplify.

The ways of worldly involvement may come in one or another of three degrees. If a natural thing is such that its variation, multiplicity, or dependence on other natural things is extrinsic to the thing itself, then its degree of metabolic, compositional, or situational involvement is *grade I*. If its invariance, unity, or independence from other natural things is extrinsic, then its degree of metabolic, compositional, or situational involvement is *grade III*. And if variation and invariance, multiplicity and unity, or dependence and independence are both intrinsic to it, then its degree of metabolic, compositional, or situational involvement is *grade II*. The three-fold involvement of continuants cannot be grade I or grade III. It must then be grade II. Continuants are essentially beings-in-becoming.

Because grade II involvement in the world is liminal, conceptually amphibious, a mean between metaphysical extremes, we are tempted to see it as a combination of grades I and III. Doing so would be a mistake, however, since splitting apart the being of a continuant would be tantamount to rendering the concept of it incoherent. We should also resist the temptation to reduce grade II to grade I or III. Reductions to grade III have been common among philosophers and scientists from the seventeenth century to the present. Descartes reduced corporeal things to nothing but matter in differential motion. Hence, properly speaking, they do not endure. They are like eddies in a stream, without any intrinsic invariance, individuality, or worldly independence.[18] In reacting against this misconstrual of individual substances, Leibniz posited metaphysical atoms, called "monads," which are everlasting, indivisible, and "windowless" entities. For monads, however, all natural change, composition, and influence from without is extrinsic. Hence their involvement in the world is grade I rather than grade II.[19] The trick is to explicate or otherwise elucidate grade II involvement without assimilating it to grade III, to grade I, or to their sum.

To make the difficulty of this task more palpable, let us turn briefly to Saul Kripke's *Naming and Necessity*. There we are told that the name "Kurt Gödel," for example, rigidly designates a singular thing which might not have been named "Kurt Gödel," might not have proven Gödel's incompleteness theorem, might not have been a mathematician, etc. As Kripke notes, his theory of reference entails some sort of essentialism with respect to individual things (as does my account of continuants), yet what sort is not clear, for we are also told that a given table T is necessarily the particular materials out of which it was made, and that Queen Elizabeth necessarily originated from the particular gametes G and G′.[20] The former speculation entails mereological essentialism for things such as T, and mereological essentialism is true only if the compositional involvement of such things is grade III. Hence table T and the like are not genuine continuants. Whether this includes such naturally occurring things as Kurt Gödel, Kripke does not say. The latter speculation is surprising, since his procedure has been to begin with some singular thing already in the actual world, ask how it might or might not have been otherwise, and to answer this question by appealing to our pretheoretical "intuitions." In the case before us, however, we are asked to consider the circumstances in which a singular thing begins to exist. Beginning to exist, like ceasing to exist, is a liminal event, and when our hunches ("intuitions") about how actual things might have been are extended to such events they cease being reliable. It is no surprise, then, that philosophers have found his necessity-of-origin speculation to be inchoate and contestable, and we do not expect the issue to be resolved any time soon.[21] In any event, Kripkean originalism does not entail that the situational involvement of Elizabeth, Blackie, and the like is grade III. For it may be that they are necessarily emergent entities, things which have come to *be*, whose subsequent three-fold involvement in the world is grade II until they cease to be.

It is easier to say what a continuant is not than to say what it is. It is not a bare or thin particular. If individual, it is neither a simple unity nor a mere heap. Its attributes, how it is, are neither altogether necessary nor altogether contingent. Its substance, what it is, is neither a this nor a such but an unanalyzable this-such. It has an inner nature or essence with which it cannot simply be identified and yet from which it cannot be separated even in thought. And, as we have just now seen, it is an entity whose grade II involvement in the world cannot be reduced to grade III or grade I or to a combination of the two. Trying to make our way through these neither-nors is like trying to find the least risky route between Skylla and Charybdis. Until we have done so, however, the concept of a continuant will not be as clear and distinct as we would like, and continuants themselves will remain epistemically as well as ontically problematic.

## EPISTEMOLOGICAL DIFFICULTIES

Realism with respect to continuants is presupposed in our everyday understanding of things, in what Husserl has called "the natural attitude." So strong indeed is this presupposition that skeptics can hardly dislodge it. And yet, on reflection, many continuant-like things do not appear to be the entities we usually take them to be.

Consider utensils. They populate and structure our world. In some sense they are real. But in what sense? Take for example my favorite pen, the one I call "My Pen." Ordinarily I regard it as a precious continuant. People around me do not share my fondness for My Pen. For them, it is just another writing instrument, yet they agree in taking it to be an enduring thing. Materialists, however, argue that My Pen is a physical object whose compositional involvement in the world is grade III rather than grade II. Functionalists argue instead that the essence of a utensil is in most cases its use. These so-called essences, however, are grounded in our practices and may change therefore as our practices change: books may become doorstops or flower-presses, hairbrushes may become back-scratchers or spankers, tables may become chairs or beds or firewood, and so forth; the possibilities are practically unlimited. But the natures of continuants do not change as our practices change; what they *are* is is not ultimately up to us. Hence utensils as such are not continuants. They are continuant-like pseudo-entities, ghosts in the "machinery" of the natural world.

The same is true of artifacts generally, as Stephen Schwartz and others have argued,[22] and of things insofar as they are matters of practical concern. All such things ancient Greeks called *pragmata*. And *pragmata* as such, Plato observed, are not entities (*onta*).[23] There is no substantial *there* there. It does not follow that such things are merely subjective. Their "essences" are no more in our heads than are the meanings of all words we use. They are objective, publicly accessible, there within the *pragmata* themselves. But it is we who sustain them in our communal practices. They are part and parcel of everyday life, *our* world. If we or our world were to wake up, as it were, all *pragmata* as such would vanish.

There is a complication. Take for example a slave called "Boy." For his master, and for the society to which he belongs, Boy is a living instrument whose essence lies in his utility to his owner. In reality, however, Boy is an adult human being whose real name, we should say, is "Ayi Kwei Appiah." Ayi Kwei Appiah is evidently an individual continuant. Hence, unlike My Pen, his involvement in the world is grade II. As Boy, however, he is a *pragma* in the world of his master. He is an entity by nature and a non-entity by convention, two things in a single body, one of which is but a shadow of the other. The same is true of his master and the master-class to which he belongs, though it is more difficult for members of this class to distinguish the realities they are from what they seem to be as denizens of their social world. Philosophers endeavor to see through collective illusions of this sort. Take for instance the new anti-Fregean theory of reference, according to which "Boy" and "Ayi Kwei Appiah" are rigid designators differing only in their associated connotations, not at all in what they denote. But for the master who named him "Boy," and the linguistic community to which he belongs, the predominant connotation so saturates the denotation that they can hardly tell the difference. A Fregean theory of reference better represents *their* reality.

It appears that something is a continuant only if it is natural. But again there is a complication. Must a continuant be naturally occurring? Imagine we have synthesized a liquid by combining two volumes of hydrogen with one of oxygen gas in a test tube and introducing a spark. Is this artificially generated stuff a portion of water? Few would deny it. Or put yourself back in the year 1936. Imagine that you and Emilio Segré have just found some artificially produced technitium, a heretofore missing chemical element which no one has yet found to be naturally occurring.[24] Should you nevertheless regard it as a continuant or as composed of continuants? Again the answer seems to be "yes." But then what is to stop us from regarding as continuants such things as kerosene, PVC and other assorted plastics, artificially propagated cultivars and clones, plants grown from genetically modified seeds, animals produced by in vitro fertilization, or subatomic particles produced in manmade accelerators? Nothing, it seems, *if* the involvement of these things in the world is grade II, *if* the manner of their genesis is incidental to what they are, *if* they are things having intrinsic natures or essences. Grade II tokens, however, are not easily distinguished from grade III tokens; emergent entities are not easily distinguished from ongoing processes of generation; and continuants do not wear their natures on their sleeves. For all we know, the liquid produced in that test tube is not a portion of water, as we and chemists believe, but something we should call "fool's water."

To make clearer still the difficulty here, consider an array of naturally occurring things that includes mountains, boulders, glaciers, forest fires, air, hurricanes, sunspots, planets, galaxies. These are continuants only if they exist independently of our ways of regarding them, and only if they have natures in virtue of which they remain the same in and through change as long as they exist. We have our hunches, of course, but hunches are not knowledge, even when they happen to agree. Even water and gold, for us exemplary continuant masses, may not be the enduring entities we take them to be. For not only are their (putative) natures inevident, recognized authorities have told us in no uncertain terms that these so-called masses are really just collections of molecules, and the compositional involvement of such collections is plainly grade III rather than grade II. What is ordinarily exemplary for us no longer appears exemplary in reality.

## EPISTEMOLOGICAL DIFFICULTIES (CONT.)

Let us begin anew with plants and animals. Ordinarily we take them to be exemplary continuants. On closer study, however, we find plenty of hard cases, that is to say, putative continuants which turn out to baffle our recognitional abilities and frustrate our desire to know. I set them out under five headings.

### *Counting Problems*

An aspen grove appears to be several locally separate trees, yet it turns out it has a single root system. Transplanting one of the trees, we find that it survives in its new, detached environment. Are there now two continuants, or the number we started with? A Portuguese man-of-war appears to be a single thing, yet it turns out to be an instance of obligate symbiosis in which genetically different cells are physiologically united. Plant chimeras, lichens, and many sponges are similarly constituted. How do we enumerate the continuants in such instances? An ant colony appears to be many individuals, yet so well organized is it that some regard the colony as a single superorganism. The pseudoplasmodium of a slime mould appears to be a single multicellular entity, but it came to be out of many unicellular organisms and sometimes disintegrates into those same organisms or their descendents.[25] Is it many or one?

### *Boundary Problems*

Blackie appears bounded by his fur coat. But its outermost electrons have no determinate locations and their electromagnetic force-fields extend indefinitely in all directions. And since Blackie is topologically a tube, we must consider the inner boundary as well, including the gut tract in which there are many bacteria whose relationship with him resembles the cases of obligate symbiosis just mentioned. Yet physics and geometry alone cannot specify his spatial limits; we must also find a way of deciding which materials are parts of him. They are, says Peter van Inwagen, those and only those now caught up in Blackie's life.[26] The food in his stomach is not part of him, nor are any dead cells or the fecal matter collecting in his large colon. Nor are many other bits of matter now streaming through his body. What about the bacteria flourishing symbiotically in his gut tract? Are they not caught up in Blackie's life, though to a lesser degree than the mitochrondria in his cells? Perhaps *being a proper part* comes in degrees, as van Inwagen implies, but if so then Blackie's spatial boundaries will be very difficult to determine. Perhaps there is no fact of the matter enabling us to tell whether these bacteria are among his bodily parts, or whether the heartwood in an oak, though no longer composed of living cells, is still caught up in its life.

*More Boundary Problems*

Plants and animals come into being and perish. That much seems clear. But when we look more closely, things go out of focus. We find ourselves unable to pinpoint, in any given case, the very moment when a continuant began to exist or when it ceases to exist. There are, it seems, times when it is exists, times when it does not yet exist or no longer exists, and the two intermediate periods in which the continuant is coming to be or perishing. These "periods" should be point-like events, yet they seem to be processes. It is very difficult, moreover, to determine the final moment of each, the moment at which this living thing finally begins or ceases to exist. And when does it begin to die? This boundary, too, seems a little vague. Where we draw all these lines may to some extent be up to us.[27]

*Metamorphosis*

A caterpillar turns into a butterfly if nothing gets in the way. The transformation of one into the other, biologists call "holometabolous." The caterpillar as a whole is destroyed, leaving only a few tiny clumps of cells, called "imaginal discs," from which the butterfly develops by using the rest of the former caterpillar as food.[28] Is this a case of nonsubstantial change, or has one continuant supplanted another? Biologists tell us their genetic structure is the same, yet it hardly follows that they are one and the same individual. Another example: Rhizocephalans (root-heads) of the genus *Sacculina* are parasitic crustaceans that begin, apparently, as free-swimming nauplius larvae. On finding a suitable host, usually a male crab, the female metamorphoses into a cyprid larva that crawls over the hard surface until she finds a weak spot, whereupon she fastens herself and metamorphoses into a dart larva that is little more than a hypodermic needle for inserting a primordium (sometimes a little slug of cells, sometimes a single cell). Once inserted, the primordium multiplies quickly and spreads root-like throughout the host, emasculating and otherwise taking complete control of him (though without, it seems, killing him). She then produces a cluster of egg-bearing sacs that protrude from his underside. Now the nearby male nauplius, having also changed into a cyprid larva, metamorphoses yet again into a hairy little mass of cells which enters this protruding cluster. There he loses the hair and becomes a sperm-producing gonad tucked away in a tiny receptacle within the female.[29] Is this egg-producing mistress of the crab identical in number with the nauplius larva from which she came? Are the male root-head and the once-male host now too dependent on their situations to be counted as continuants? The distinction between substantial and nonsubstantial change is here very difficult if not impossible to make.

*Intermittent Existence*

Rotifers and tardigrades appear to be individual continuants. Biologists have placed these tiny animals in baths of liquid helium, and then warmed them back up. While immersed in the helium their body temperatures were close to absolute zero; almost no molecular activity of any sort was occurring in them. They must then have been dead, it seems, and yet on reheating they revived.[30] Call a tardigrade we are about to freeze "Lazarus," and his successor, the one moving about after reheating, "Lazarus*." To find out whether Lazarus and Lazarus* are the same in number, not merely in kind, we need to know more about the thing Lazarus became when cooled to a few degrees above absolute zero. Call this thing "X." Biologists say that X is ametabolic, hence not alive; that X is identical with Lazarus in a cryptobiotic state; and that Lazarus* is therefore identical with Lazarus. But is the distinction between biotic and cryptobiotic states as sharp as they suppose? And supposing it is, what then is the ontic status of X? To be in a cryptobiotic state is to be only potentially alive. A detached finger is a finger in name only, Aristotle said, unless it can function as a finger when reattached, in which case it is still potentially a finger. To be potentially but not actually a finger is not to be a finger simply. X is not actually Lazarus*. Nor is it Lazarus. Nor is it a mere corpse. Lazarus and Lazarus* are not evidently identical or evidently diverse.

Hard cases of the first, fourth, and fifth sorts may strike many of us as exceptional, yet one should not underestimate the number of such exceptions: living things are far more varied and wonderful than we usually imagine. And hard cases of the second and third sorts are pervasive. Blackie and the like are genuine continuants if they have natures, or if they are intrinsically invariant as well as varying, individual as well as dividual, and independent of as well as dependent on their environments; but knowing this much does us little good when it comes to discovering whether Blackie or the like is the entity it seems to be. Reflecting on these cases undermines one's commonsensical acceptance of realism with respect to continuants. Indeed, we now see that common sense may not be reliable in these matters, since it is predominantly pragmatic and liable therefore to mistake continuant-like *pragmata* for enduring entities.

As the plausibility of continuant-antirealism looms large, so too does the thought that continuants themselves are indefinite or vague. Let us explore this thought before trying to decide whether realism with respect to continuants is, after all, an acceptable position.

## THE TROUBLE WITH MATTER

Philosophers tend to agree that continuants are vague in virtue of their materiality. For this very reason, some have argued, there can be no continuants. Their argument goes something like this:

1. All vagueness is epistemic or semantic rather than ontic.
2. Therefore, if the concept of something ontic requires that it be inherently vague, that concept should be expunged from one's ontology.
3. The concept of a continuant requires that any such entity be inherently vague.
4. Therefore, the concept of a continuant should be expunged from one's ontology.

To this argument continuant-realists have responded by accepting premise 3 and rejecting premises 1 and 2. Van Inwagen, in particular, has endeavored to show that ontic vagueness is admissible.[31] For my part, I agree that premise 1 by itself is a bare assertion whose truth-value is far from obvious. Yet suppose it is true. So too, then, is premise 2; and the key question concerns premise 3. I agree that continuants seem intrinsically vague in virtue of their materiality. But seeming vagueness is not necessarily real. Premise 3 may not be true. It stands in need of an argument. Continuant-antirealists have offered several. Let us examine three of them.

One is a Sorities-type reductio argument that builds on hard cases of the second sort. Let there be a continuant *a*. It is necessarily embodied and thus spatially delimited. There is the region definitely within its body (call it "IN") and the region definitely outside its body (call it "OUT"). And since *a* is essentially material, the boundary between IN and OUT is indefinite; there must be an intermediate region in which bits of matter are neither definitely within nor definitely without the body of *a*. Call this liminal region "LIM." The boundary between IN and LIM, or between OUT and LIM, cannot be sharp, since *a* is intrinsically vague. Now, beginning anywhere in IN, move straight outward in any direction, particle by particle, until you are about to enter LIM. Move outward by one more particle in the same direction: you are still in IN, since otherwise the boundary between IN and LIM would be sharp. Take one more step in the same direction: same result. And so on. No matter how far you travel in any direction, you remain within *a*, which is absurd. Alternatively, beginning anywhere in OUT, move straight inward: by the same stepwise argument you find that you can never reach IN, which is also absurd.

A second Sorities-type reductio argument, made by Mark Heller and others, builds on hard cases of the third sort. Let there be a continuant *a* having finitely many smallest parts. We remove one particle at a time until *a* has definitely ceased to exist. There is the period of time when it definitely exists (call it "E") and the period of time when it definitely will not exist (call it "NE"). Since *a* is essentially material, the boundary between E and NE is indefinite. There will be an interval of existential indefiniteness, when *a* does not definitely exist or definitely not exist. Call this interval "IND." The boundary between E and IND cannot be sharp, since *a* is intrinsically vague. Now, beginning sometime in E, keep removing particles, one at a time until you are about to enter IND. Remove one more particle: you are still in E, since otherwise the boundary between E and IND would be sharp. Remove one more particle: same result. You may continue this process until there are no more parts, and at each step *a* still definitely exists, which is absurd. Continuant-realism is again shown to be necessarily false.[32]

A third reductio argument, offered by Panayot Butchvarov, builds on problems we have in trying to keep track of alleged continuants. Blackie is an enduring material being. But matter itself is indefinite; there is no way of telling whether "it" endures or fails to endure. Hence no material thing as such is identifiable over time. To be an endurant, then, Blackie must not be essentially material. But he is. Therefore, if he exists, he cannot be the sort of thing that endures; he cannot be a continuant. We may *claim* that he is, of course, but there is no basis in reality for thinking that *a* now is the same individual as *a** a moment ago, or for thinking that *a* and *a** have the same nature or that they have different natures. Our claim would nothing more than an idle speculation.[33]

These arguments are not altogether new. Eleatics, Megarians, and modern empiricists have offered similar reductios. Their targets have included the possibility of change, potentiality, matter, substance, and causality—in short, the subject matter of natural philosophy.[34] Their conclusions in some instances entail antirealism with respect to nature, which contradicts our hypothesis H2. But H2 is compatible with the conclusion that no continuants could exist, so we must find some additional reason for dismissing it, or else give up the thought that there could be natural kinds in the primary sense of term.

All three arguments assume that continuants are essentially material and therefore intrinsically vague. The second conjunct seems acceptable.

I agree, moreover, that if the first were true, a continuant's compositional and metabolic involvement in the world would be grade III, and that with respect to grade III objects these arguments are conclusive.[35] Continuants, however, are grade II entities. Hence the essence or nature ($n_x$) of such a thing ($x$) is not its matter. What then could it be? Consider Blackie. He is, I suggested, a living thing whose material parts are (more or less) caught up in a particular life-activity. This life-activity, I now suggest, is his nature $n_x$, what he *is*. It does not follow that his materiality is incidental to him. Indeed, were that so, his involvement in the world would be grade I rather than grade II. His materiality must then be necessary without being of the essence. This distinction is admittedly hard to understand, yet no one to my knowledge has shown it to be unintelligible.

One may say that a continuant $x$ is intrinsically invariant, unitary, and situation-independent in virtue of its nature $n_x$, and intrinsically varying, multiple, and situation-dependent in virtue of its matter $m_x$. But it does not follow that $x$ is essentially $n_x$ plus $m_x$, or that $n_x$ and $m_x$ are metaphysically on a par.[36] It may also be that the necessary materiality of $x$ entails some intrinsic vagueness. Perhaps $x$ has a soft core. But again, if the concept of a continuant requires that such entities be somewhat vague (premise 3), it remains to be shown that vagueness is never ontic (premise 1), or that something exists only if it is identifiable, an assumption I questioned in my examination of protometaphysical antirealism.[37] It may also be that premise 3 is not true. Continuants may seem indefinite as well as definite because our conception of them is inadequate. But if so, the vagueness is epistemic rather than ontic, and premise 1 is not yet violated. Of course, the fact that for many centuries ontologists have tried without complete success to clarify the concept of a continuant makes us suspect that the problem lies in the things themselves. But we have not yet found this to be the case. For all we know, the problem lies in us, or perhaps in the concept itself. In any case, it has not been shown that continuant realism is necessarily false.

Is it in fact true? But first, a brief comment on what continuant antirealists now take to be the main alternative, namely, realism with respect to four-dimensional occurrents. Continuants have no temporal parts. As entities, they are complete at every moment of their existence. Each endures—i.e., $x$ and $y$ at different times are the same entity in different states—until it ceases to be. Four-dimensional occurrents do have temporal parts, and are complete only when their final parts come into being. Each perdures—i.e., $x$ and $y$ at different times are diverse parts of the same entity—until there are for it no succeeding temporal parts. The difference between endurants and perdurants is huge. But is it the case, as Mark Heller and others claim, that material perdurants are not inherently vague, whereas endurants are? I think not, for two reasons. First, they too are material. Therefore, if the very materiality of continuants makes identification impossible, the same is true of these perdurants.[38]

Second, to individuate perdurants we begin with things such a Blackie (or if you wish, a wooden table), and reconceptualize them as four-dimensional objects.[39] Let some given entity be conceptualized as the endurant *a*. A pair of Sorites-type arguments were produced to show that *a* is spatially and temporally vague. Now reconceptualize this entity as the perdurant *a*′. We can produce an analogous pair of Sorites-type arguments showing that *a*′ too is spatially and temporally vague: any one of its temporal parts has spatial regions IN, OUT, and LIM, etc; and its succeeding temporal parts are such that some definitely belong to to the not yet completed *a*′, some definitely come later than the completion of *a*′, and some belong to an indefinite interval when *a*′ is neither incomplete nor already complete, etc. Therefore, by parity of reasoning, endurant *a* is vague if and only if perdurant *a*′ is vague. But again, such arguments are not evidently sound. It has not been shown that perdurant realism *or* continuant realism is necessarily false. The question before us now is whether the latter is true.

## A FIFTH HYPOTHESIS

The assumption that continuants are among the things that exist is part and parcel of our natural attitude toward the world. In reflecting on this assumption, problems aplenty have come to light. Now it would be dogmatic for us to regard continuant realism as true without first weighing the reasons *pro* and *con*.

An argument in favor goes as follows. "Suppose continuant antirealism is correct. Then those who advocate it are not continuants. Yet they seek to understand and modify the world around them. They think, argue, and change their minds. They have interests and are able to change and be changed. What sort of beings are these agent-patients? Continuants, of course, so our original supposition must be incorrect."[40] But this reductio will not do, for at least two reasons. First, it remains to be shown that the beings in question are the agent-patients they appear to be. Second, while continuant realism may figure in our best explanation of such phenomena, this explanation may not be the one we seek and may even impede our search inasmuch as it tacitly receives much of its force from common sense.

Another argument in favor goes as follows. "In pre-reflective experience we directly perceive not sense data or mere sensory manifolds, as empiricists used to believe, but individuals such as Blackie and portions of stuff such as water. This, surely, is the non-inferential perceptual basis of our belief that continuants exist." But this argument, too, will not do. For granting that we do directly perceive Blackie or the like, what we then perceive is a continuant-like object which or may not be a continuant. It is the real thing, we have seen, only if it has some inner nature in virtue of which it *is* what it is. But the nature of any given continuant is occult. We can have no direct perception of *it*. Hence, for all we know, none of the continuant-like objects we perceive are the entities we take some of them to be.

To weigh this skeptical rejoinder, recall the discussion of universals and particulars in chapter 3. The properties of substantial particulars, I said, may also be instances of universals. Suppose they are. Then these properties are manifestations both of the universals they instantiate and of the substance they modify. They are, as it were, an epistemic gateway between two fundamentally different domains. By looking through them in one direction, I suggested, we may in time become directly aware of those universals. And by looking in the opposite direction, I now suggest, we may in time become directly aware of what this thing is. In other words, as we can see the universals epistemically present in their instances, so too can we see the *what* that is epistemically present in the *how*. Even in the perception of a continuant-like object, we observed, it must appear to have an evidently inevident nature; otherwise it would not even be continuant-*like*. The problem, obviously, is that the nature it appears to have may not be the nature or essence of a continuant; there may in fact be no genuine $n_x$, no substantial *there* there. But from this it does not follow that direct awareness of that nature is impossible. Similarly, from the fact that universals are evidently inevident in a thing's perceived properties—that here too we may be fooled—it does not follow that direct awareness of instantiated universals is impossible. Underestimating our cognitive abilities, as skeptics are wont to do, is no less an error than overestimating them, as dogmatists are wont to do. To avoid the former error, it may help to consider such everyday events as seeing anger, shame, or elation in a person's countenance; what people want in how they behave; agency in actions; irritability in irritation, malleability in malleation, and so on. As Husserl, Evan Fales, and others have observed, seeing-in is a common occurrence.[41] Our lives depend on it. True, it is not always veridical. But to require that seeing-in be transparently veridical if it is to have any cognitive value at all would be to disallow *a priori* all direct cognitive access to the world.

A skeptic may reply as follows. "Let it be granted that the core of a continuant is visible in its surfaces. Still, any insight into it that we may have depends not only on how we perceive these surfaces, many of which are fleeting, but also on how we sortally or otherwise typify the continuant itself. Thus will our awareness of its core always be colored by our ways of singling out and (re)identifying it. And we have little reason to suppose that these are grounded in the things themselves rather than in us."

It may be true that we gain no insight into the nature or essence of a continuant without sortalizing it. But sortal dependence does not entail sortal relativity, any more than the discovery of some truth must be relative to the inquiry on which this discovery depends, or than skiing well must involve following rules of the sort we followed when learning to ski. We get a fix on continuants by trying to typify them as substances; yet with practice, by paying close attention, we can become expert continuant watchers whose insights are sometimes free of these typifications.[42] It may also be true that we gain no insight into the nature of a continuant without being perceptually aware of the ever-changing qualitative manifold it presents. But again, that insight depends on sensory awareness does not entail that it is relative to such awareness or that it is really an observation-based inference. What we need is a theory of direct recognition analogous to Kripke's theory of direct reference, according to which the proper name of an individual designates it rigidly, i.e., so anchors us semantically in what the named individual *is* that we can identify it across different possible scenarios. Similarly, the direct recognition of this individual so anchors us cognitively in its essence or nature that we can keep track of it as it changes over time. One may doubt that Kripke's theory of rigid designation is superior to the Fregean alternative, or that my theory of nature-recognition (which I have barely outlined) would prove superior to empiricistic alternatives.[43] But then one should also doubt these alternatives. There is a burden of proof on both sides. And the proofs we seek are hard to come by, since the scope and limits of our cognitive abilities are hardly evident to our cognitive abilities.

Given this double-edged uncertainty, it would be dogmatic to insist that continuant realism, or continuant antirealism, is correct. Yet nothing prevents us from accepting either as a hypothesis. And given our remarkable ability to single out and track some continuant-like objects, an ability that skeptics have not shown to be spurious, it seems reasonable to accept tentatively the realist alternative. Let us do so.

H5. There are continuants.

This hypothesis, together with hypotheses H1–H2 and H4,[44] gives us most of what we need to begin answering our questions about natural kinds. Notice, however, that just as H4 did not tell us which putative universals are genuine, so H5 does not tell us which (if any) of the continuant-like objects we observe are genuine continuants. Notice, too, that the epistemological and metaphysical difficulties beclouding H5 seem even greater than those surrounding H4. And notice that, as we seek a middle way between metaphysical outreach and epistemological rigor,[45] the very resource on which we now rely—namely, our everyday understanding of the world—appears less and less reliable. We are proceeding with cautious daring amid looming problems as the ground beneath our feet begins to deliquesce. Welcome to natural philosophy.

## NOTES

1. Cf. Atran and Medin 2008. See also A. B. Comstock, *Handbook of Nature Study* (Ithaca, NY: Cornell University Press, 1987), Part I.

2. A published work on natural kinds that exemplifies what I have in mind here is Wilkerson 1995. Wilkerson is, he says, an avid gardener.

3. Aristotle, *Metaphysics* IV.4.

4. Cf. P. T. Geach, *God and Soul* (London: Routledge & Kegan Paul, 1969), 70–72. Enduring such changes is not real endurance. A universal "undergoes" only Cambridge changes as its instantiations come and go. It is not a continuant.

5. I say "to some extent" in connection with the last-mentioned properties because the length or mass of a body cannot be less than its rest-length or rest-mass; and if its rest-shape is spherical, say, there can be no reference frame in which it is dodecahedral; and two events cannot be simultaneous if they do not fall within the same "light cone."

6. Hume 2002, I.iv.6–7.

7. Consider Quine 1960, 199f.

8. Again, these attributes may be relational as well as monadic. They appear also to be irreducibly dispositional as well as categorical: continuants are powerful particulars.

9. Cf. Almog 1991, 230ff. Like Almog, I focus on things in the actual world. Unlike him, I also focus on variation and invariance in such things.

10. A further complication: An attribute that is contingent for $x$ at one stage may be necessary at another. Blackie, if male, cannot become a female; but it does not follow that he could never have become one. Cf. Mellor 1977, 301; and Almog 1991, 235n5.

11. Regarding it as two would commit one to the Bradley Regress, yet a this-such is not simply one either.

12. Blackie also *has* various attributes, but we would not say that this particular substance *is* any or all of them. That my account here is consistent with our pre-philosophical notion of essence: Gelman and Hirschfeld 1999, 403–46.

13. Cf. Butchvarov 1979, 122–53.

14. Wiggins 1980, 47, 58–59, 68.

15. The unity and multiplicity could be that of an individual such as Blackie or of a mass such as water. Both, presumably, are continuants. Universals and tropes are not; there is no distinction to be made between such entities and their natures or essences.

16. Thus, while the particular substance $x$ is ultimate in the order of categorization, its nature $n_x$ is ultimate in the order of metaphysical explanation. This difference becomes manifest when one compares Aristotle's *Categories* 2ff with the books VII–IX of his *Metaphysics*.

17. Harré and Madden take continuants to be powerful particulars, and initially identify their natures with causal powers or dispositions. In their subsequent account, however, they implicitly reject any such identification, for they say that natures and dispositions are interwoven, tied together, complementary, and that the latter manifest or express the former, not conversely. Cf. Harré and Madden 1975, 6, 11, 13, 53, 82, 97.

18. See chapters 6–7 of this volume.

19. Consider also the teachings of those who claim that each of us is really an immortal soul whose immersion in the vicissitudes and materiality of the world abroad is incidental to it. Leibniz himself noted the similarity between monads as he conceives them and angels as conceived in Scholastic theology.

20. Kripke 1980, 110–15.

21. Cf., e.g., McGinn 1976, 127–35; Noonan 1983, 1–20; Forbes 1985, 133–45; Mackie 2006, 28f, 93–117.

22. Cf. Schwartz 1978, 566–74, and his 1985, 182–95. Also Wiggins 2001, 35–43, 86f, 89–102, 133–36; and Wilkerson 1995, 99, 105–07.

23. Plato, e.g., *Cratylus* 385e6–386a3, *Theaetetus* 152a1–8. Translated into Latin terms: Of the *res* we deal with in everyday life, many are *facta* rather than *entia*.

24. Scientists will soon find traces of its several isotopes in terrestrial fission products of uranium and plutonium, and evidence of it existence in the light emitted by stars called "S-type red giants."

25. Cf. S. L. Stephenson and H. Stempen, *Myxomycetes: A Handbook of Slime Molds* (Portland, OR: Timber Press, 1994), 20 and context; J. T. Bonner, *The Social Amoebae* (Princeton, NJ: Princeton University Press, 2009), 24–33.

26. Inwagen 1990, 82–96, 121, 170, 175; also Wiggins 1980, vii, 70, 86–89, 175, 182, 207.

27. It was customary in ancient Greece for fathers to name their children soon after they were born. Aristotle probably got his name in some such naming ceremony. Perhaps at that time he had already existed for several months. Or perhaps the living thing so named was still only potentially the individual continuant we call "Aristotle," who may not have come into being for several more years. And perhaps this continuant ceased to be several days before something still called "Aristotle" died and was buried. Might not all our proper names for human beings and other living things be like this?

28. Cf. Gilbert 1994, 729–35; Inwagen 1990, 155; and for an evolutionary account, A. J. Yang, "Modularity, Evolvability, and Adaptive Radiation: A Comparison of Hemi- and Holometabolous Insects," *Evolution and Development* 3 (2001): 59–72.

29. Cf. F. R. Schram, *Crustacea* (New York: Oxford University Press, 1986), 473–78; J. T. Høeg and J. Lutzen, "Life Cycle and Reproduction in the Cirripedia Rhizocephala," *Annual Review of Oceanography and Marine Biology* 33 (1995): 427–85; and S. J. Gould, "Triumph of the Root-Heads," *Natural History* (July 1996): 10–16, reprinted in his *Leonardo's Mountain of Clams and the Diet of Worms* (New York: Harmony Books, 1998).

30. Cf. Keilin 1959, 149–91; R. C. Brusca and G. J. Brusca, *Invertebrates* (Sunderland, Mass.: Sinauer, 1990), 675f; J. S. Clegg, "Cryptobiosis: a Peculiar State of Biological Organization," *Comparative Biochemistry and Physiology* ser. B 128 (2001): 613–24; D. A. Wharton, *Life at the Limits: Organisms in Extreme Environments* (New York: Cambridge University Press, 2002), 6–8, 13, 180 and context, 212, 246f, 277; L. Rebecchi, T. Altiero, G. Guidetti, "Anhydrobiosis: the Extreme Limit of Dessication Tolerance," *Invertebrate Survival Journal* 4 (2007): 65–81. Biologists seem to think that Lazarus in a cryptobiotic state is like a machine that has been turned off but is still ready to go; it merely needs an input of energy it can use.

31. Inwagen 1990, 217ff.

32. Unger 1979, 117–54; Heller 1988, 109–31, his 1990, 68–109, and his 1992, 695–704; Horgan, 1994, 173ff; Williamson 1994, 111ff; Sider 1997, 197–231.

33. Butchvarov 1979, 161ff; Robinson 1982, chapter 17. Compare Hirsch 1976a, 46–55, 65, and 1976b, 376–89.

34. I am unaware of any comprehensive account of this thread in the history of philosophy. On the history of Sorites-type arguments, cf. Williamson 1994, 8–69. The first to propose such an argument appears to have been Eubulides, a Megarian. The first to argue that matter itself is unintelligible appears to have been Berkeley, though one finds antecedents in the writings of Aristotle and others.

35. M. Heller seems to make no distinction between continuants and *pragmata*: Heller 1990, 30f, 45–52, 69f. Not unrelated is the fact that his paradigmatic continuant is a wooden table.

36. P. Butchvarov seems to to think that the essence of a material thing would have to be $m_x$ or $m_x$ plus $n_x$: Butchvarov 1979, 155, 164, 168,172, 232. This is tantamount to denying the possibility of grade II involvement in the world.

37. Notice that, outside of formal set theory, it is difficult to state precisely the relevant criteria of identity. There seems to be some tension, then, between insisting on identifiability and insisting that one's subject matter be real or natural.

38. On the difference between endurants and perdurants: Chen 2003, S962–S974. Mark Heller says very little about the matter of his four-dimensional "hunks of mat-

ter." Perdurantists could appeal to general relativity theory, in which matter has been geometrized, but then they would have to admit that quantum indeterminacy prevails in very small regions of space-time; cf. Greene 1999, 128 and context.

39. To my knowledge, no perdurantist has tried to derive four-dimensional individuals directly out of space-time physics without any appeal to our pre-theoretical understanding of things. And it is hard to see how any such attempt could succeed.

40. For similar arguments, cf. Butchvarov 1998, 151–52 with 96–110; and W. R. Carter, review of Heller 1990, in *Philosophical Review* 102 (1993): 125. Notice that Kant, in his defense of continuant realism (within his protometaphysical antirealism), emphasizes action and reaction as distinguished from mere commotion: Kant 1996, B249–50.

41. Cf. Husserl 1970, I, 332–44 and II, 441–47, 773–815; A. R. White, *Attention* (Oxford: Blackwell, 1964), 69–70, 116; Fales 1996, xii, 87–89, 137, 140; and P. Valéry, *Eupalinos ou L'Architecte*, in *Œuvre* II (Paris: Gallimard, 1960), 99. That we are thus directly aware of causes, see: W. James, *Essays in Radical Empiricism* (New York: Longmans, Green and Co., 1938), 185–86; Harré and Madden 1975, 49–62; A. Flew, "Hume and Physical Necessity," *Iyyun, The Jerusalem Philosophical Quarterly* 39 (1990): 256. I use "see" metaphorically, in keeping with our ordinary ways of speaking. Sometimes we use the word "feel" analogously, as in, "Do you not feel the springiness (elasticity) of the rubber ball you are squeezing in your hand?" Dispositions of this sort persist in their expressions. Others do not—fragility, for instance.

42. The difference between sortal-dependence and sortal-relativity is a major theme in Wiggins 1980. See also Fales 1996, 51, 90, 96, 105–09; Butchvarov 1998, 27; and Searle, *Intentionality: An Essay in the Philosophy of Mind* (New York: Cambridge University Press, 1983), 120ff. On the role of attention, see: Gurwitsch 1964, 4–5, 307ff; and Fales 1996, 63, 106. For a discussion of pattern recognition that supports this point, see: W. G. Chase and H. Simon, "Perception in Chess," *Cognitive Psychology* 4 (1973): 55–81.

43. These alternatives are now commonly regarded as forms of "internalism," whereas the theory I propose is a form of "externalism." But for reasons given in my account of the onto-logical difference in chapter 2, I prefer not to use these spatial metaphors. I also do not accept a causal theory of perception or recognition or insight, whereas most externalists do.

44. See chapters 2 and 3 of this volume.

45. Cf. Swoyer 2005, 7, 26, 32. In Weinberg 1992, 74, Steven Weinberg writes that the great mathematical physicist David Bohm was once asked, "What is complementary to *Wahrheit* [truth] in Bohr's sense of the term?" "*Klarheit* [clarity]," he replied.

# FIVE

# Natural Kinds

Chapter 1 showed how everyday experience can motivate our two leading questions, namely, "What is a natural kind?" and, "Do any such kinds exist?" Chapter 2 examined and set aside two philosophical objections to my trying to answer these questions in accordance with everyday experience. Chapter 3 showed that natural kinds exist as types only if universals exist, and that tokens of such types exist only if there are enduring things or continuants. Chapter 4 clarified and sharpened our everyday notion of a continuant, and gave reasons for thinking that individual continuants exist. Now at last, after all this preparation, we are ready to begin answering our two leading questions. In this chapter I first say what it is to be natural kind (in the primary sense of the term); and then give reasons for thinking that our everyday experience and understanding of things is insufficient to the task of answering our second question, that to answer this more empirical question with anything approaching certitude we must make use of scientific theory as well.

## WHAT IT IS TO BE A NATURAL KIND

A natural kind, I say, is a class whose members are all and only those continuants that exemplify, in virtue of their essences or natures, a single type. Alternatively, a natural kind is a type whose tokens are all and only those continuants that exemplify it in virtue of their essences or natures, and that constitute, therefore, a single natural class.

Though not as perspicuous as we might like, the terms of this answer are now clear enough that we are no longer liable to misconstrue or be utterly baffled by the answer as a whole. That is one point in its favor. Another is that it preserves and explains the ambiguity, noted early in chapter 3, between natural kinds as classes and as types. Extensionalists conceive them in the former way, typologists in the latter as well. Both are right: the concept of a natural kind has its proper employment in thought that is at once classificatory and typological. A third point in its favor is that makes the essentialism sometimes ascribed to natural kinds derivative upon the essences of its members as individual entities. A fourth, related point is that it preserves the betwixt-and-between character of a natural kind by requiring that its members or tokens be those conceptually amphibious beings-in-becoming we call "continuants." A universal is a natural kind, then, only if its tokens are natural things of this sort; and a class is a natural kind only if its members are of this sort and only if it can gain or lose members without ceasing to be. A fifth point in its favor is that it elucidates the distinctive and irreducible role this concept has in our thought.

One may object that my answer is incomplete since it leaves out natural kinds of stuff. I do seem to have assumed that the members of a natural kind are individuals, even though in earlier chapters I granted that continuants may be masses as well or instead. Let water and gold be continuant masses, portions of which are scattered about. Are they not two in kind? And are not these kinds natural-kind types? So it seems. Let my answer be complicated accordingly. Notice, however, that they are not exemplary natural kinds. For the mass-type *water* is multiply exemplifiable only in the sense that its tokens are numerically diverse portions of one and the same stuff, not in the sense that they are diverse continuants.[1] It would seem to follow that the natural-kind *water* is really a singleton class whose sole member is one of a kind. But this too is not accurate, for mass terms as such have no plural forms and hence no genuine singular forms; they do not designate *objects*.[2] Therefore, strictly speaking, water does not constitute a singleton class. Nor is it a class having two or more objects as members in virtue of their natures. It is not a natural-kind *class* at all. Yet water and gold do seem to exemplify different continuant types, so perhaps we should still count them as natural kinds. Yet the concept of a natural kind does not straightfowardly apply to them. This is why I restricted my answer as I did. It is also why Thales, a monistic mass ontologist who claimed that all things are water, would probably have been puzzled had someone asked him whether water is one of a kind.

One may also object that my answer is incomplete since it fails to specify the role of this concept in scientific explanation and prediction. It is true that the nature (essence) of a continuant explains to some extent its structure, features, and behavior; and true, therefore, that if several continuants belong to a single natural kind, they resemble one another in their structures, features, and behavior. So, for example, knowledge of what lions *are* helps us to explain the fact that they are carnivorous quadrupeds and to predict that they will hunt gazelles (if available), roar on occasion, etc.[3] From this, however, it does not follow that explanations or predictions grounded in natural kinds must also be grounded scientifically in physical laws. To *his* natural-kinds realism Aristotle joined a naturalistic teleology. This union was not necessary: one can have a teleological conception of the world without natural kinds, as in Stoic physics, and natural kinds without teleology, as in today's scientific essentialism. Nor is it necessary to unite natural kinds with laws of nature: one can have a nomological conception of the world without natural kinds, as in Galilean science, and natural kinds without laws of nature, as in Aristotelian physics. The concepts of a natural kind, a law of nature, and final causality are mutually independent. It would be a mistake, therefore, to define the first in terms of the second or third, or the second or third in terms of the first. Nothing in the concept of a natural kind tells us that explanations and predictions grounded in such kinds are going to be nomological, teleological, or something else altogether.

A third possible objection is that I have singled out one of many ways in which the term "natural kind" is used and made it the sole use. Yes, there are natural kinds of enduring individuals and stuff; but there are also natural kinds of systems whose involvement in the world is only grade III, as well as natural kinds of states, events, processes, stages of development, forces, fields, and so forth. There are at least as many concepts of a natural kind as there are different theoretical interests or disciplines. My monistic account of it is therefore unduly restrictive. But this is not correct. It was John Venn (1834–1923), closely following John Stuart Mill (1806–1873), who introduced the term "natural kind" into our discourse. Mill had set out to reform Aristotelian logic. He began by replacing the traditional nine or so categories of being with just two, namely, subjects and classes. Classes he then divided into two sorts, manmade and real. The members of a manmade class have only one or a few distinguishing properties in common; or, if they have many such properties, these are all interrelated by some discoverable law. The class of red things is manmade, as are the classes of women and of white men. The class of human beings on the other hand is real, as are the classes of animals, of plants, and of minerals such as phosphorus and sulfur.[4] What Mill called "real classes," Venn called "natural kinds," and as members of such kinds he too mentioned human beings, animals, and plants—all of which appear to be exemplary continuants.[5] Aristotle used very similar examples in his *Categories*, when trying to make clear what he meant by *ousia* (entity, substance), the primary category of being. What Mill called "subjects" Aristotle called "primary *ousiai*"; and what Venn called a "natural kind" he called an *eidos* ("species") or "secondary *ousia*." The words are different but the concepts are the same.[6] This is no mere coincidence, since Aristotle's logic and Mill's revision of it are both rooted in our everyday understanding of things. It is here, in the understanding we all share, that both the concept of a natural kind and realism with respect to natural kinds have their native home; and it is from this source that the two great traditions of natural kinds have sprung. Thus, according to Aristotelians and Millians alike, the concept of a natural kind is pretty much as I have described it. The Aristotelian tradition, already in its twilight in the late sixteenth century, disappeared altogether with the advent of Galilean science. One may agree with Ian Hacking that the Millian tradition of natural kinds is now also in its twilight, citing as evidence for this claim the fact that philosophical naturalists today today want to free natural-kinds realism from all essentialism and make the term "natural kind" flexible enough to accommodate our diverse theoretical interests.[7] But from this it follows not that our monistic conception is unduly restrictive, but that the concept of a natural kind no longer seems to have the theoretical use it was once thought to have. There is in fact only one such concept, not many, and it is of kinds whose members (or tokens) are individual continuants. Pluralistic conceptions of it are unduly lax.

I agree, however, that the term "natural kind" may be used in more than one way. For example, we may find that planets are sufficiently continuant-like, and sufficiently like one another in what we take to be their natures, that we are justified in speaking of *the planet* as a natural kind, and in claiming that our refusal to classify Pluto as a solar planet is not "merely semantic."[8] We may also find that natural kinds are sufficiently continuant-like, when considered as changeable classes, and sufficiently similar in terms of the immutable types we take to be their essences, that we are justified in speaking of *the natural kind* as a second-order natural kind, one whose members include *the squirrel, water, the planet,* and perhaps even *mass, charge, spin,* and the like. But all these classes or types would be natural kinds only derivatively, by analogical extension from the primary sense of the term. Each would be justifiable, moreover, just in case it could be shown that the proposed analogy is an illuminating one. The *term* "natural kind" may then come to have an irreducible plurality of meanings. If so, it is yet another case of focal homonymy,[9] or of family resemblance in which the family head is determined by the concept of a natural kind as I have described it and as Aristotle, Mill, and others have traditionally understood it.

To John Dupré and others, even this mild ordering of the ways in which we use the term "natural kind" will feel like a straitjacket from which they long to escape.[10] Accepting it, though, has a couple of advantages. First, it helps us to avoid linguistic promiscuity. Suppose, for example, that philosophers begin to call the class of edible fruits a natural kind. I deny that it is, on the ground that edible fruits are not sufficiently like a natural kind in the primary sense of the term. They respond by denying that there is any such family head. But if so, then apart from our ever-changing mores and proliferating theoretical disciplines, there would appear to be no restriction at all on the ways in which we use the term. In choosing to call the class of edible fruits a natural kind, or not, nothing genuinely metaphysical would be at stake. That we disagree strongly about it might be of interest to psychologists or sociologists, lexicographers or historians, but not to ontologists.

Second, by staying focused on natural kinds in the primary sense of the term we can better avoid conceptual confusion. Imagine that a philosopher denies the existence of continuants and continuant-types. The only entities, in her view, are states of affairs involving thin particulars and physical properties, which together with second-order universals constitute a spatiotemporal system governed by physical laws. These states of affairs are themselves particular, yet they fall into classes by virtue of the different properties they involve—specific quantities of mass, distance, time, charge, spin, and so forth. Such classes, or the types their members exemplify, she calls "natural kinds." We wonder why. Perhaps she is only speaking colloquially, as an anti-essentialist does when using the word "essence" to describe a perfume, or an anti-substantialist when calling David Armstrong's body of work "substantial." Or perhaps, like Descartes, she deems it prudent as well as ingenious to disguise the novelty of her thought by clothing it in old-fashioned verbiage. If so, we have little reason to suspect conceptual confusion. Not so, however, if she takes these to be natural kinds even though there are no natural kinds in the primary sense. Then we should like to know why it would not be better to give up this term and speak instead of state-of-affairs particulars and state-of-affairs types. Doing so may prove awkward at first, but it has the compensating advantage that we no longer beguile ourselves into believing that we are still employing the concept of a natural kind. We would realize that a different concept requires elucidation—that of a state-of-affairs type—and it is realism with respect types of *this* sort that is at issue. Are they universals? If so, do they differ in kind from property-universals? Or are they not universals at all, but rather concepts we employ in the classification of particular states of affairs? If so, do these particulars fall naturally into fairly distinct classes in virtue of the different properties they involve? And what is it to be a physical law? Do such laws exist? These would be our questions, and not the deceptively similar questions guiding our present inquiry.

A fourth possible objection concerns the sort of essentialism implicit in my account of those entities which may or may not belong to natural kinds in the primary sense. As possible-worlds theorists and others have pointed out, one should distinguish clearly between what some individual might have *been* (timelessly) and what it might have *become* (in time).[11] What that individual is essentially comes into focus when one considers it in the former and not the latter way. In my account of what a continuant is in its nature, I failed to distinguish sufficiently between being and becoming. Consequently the essence ("nature") of this entity has remained out of focus. And since exemplary natural kinds (if any) have entities of this sort as their members, my account of them is going to exhibit the same ontological blurriness.

I agree that what might have been should be distinguished from what might have come to be. There may well have been a time when Blackie might have been female and grey even though he could not become female or grey. I also agree that when considering continuants as entities one must proceed broadly logically (*logikôs*) as well as naturalistically (*phusikôs*).[12] My account in chapter 4 proceeded accordingly. But I also maintained that continuants are entities whose being (*ousia*) cannot be considered apart from their becoming (*genesis*), and that doing so would be tantamount to considering them as if their three-fold involvement in the world was grade I rather than grade II. This, I suggest, is what Kripke and other possible-worlds theorists have done with respect to Kurt Gödel, Queen Elizabeth, Socrates, and the like; they have turned their natures into Leibnizian-like monads, as it were, and then tried to figure out what belongs to them essentially. The results have been a muddle. Might Gödel have been named "Dick Button"? Of course. Might he have been incompetent in mathematics? Yes. Might he have been an alligator? Maybe. What about an iris? An $H_2O$ molecule? A paper clip? There is a line to be drawn here, we feel, yet we seem to have no principled way of drawing it.[13] By proceeding logically alone, by freeing ourselves from the rough ground of the natural world, our thinking has become untethered from the continuants in question. Proceeding naturalistically, as scientific essentialists have emphasized, helps us to keep our feet on the ground.[14] And in the case of continuants such as Gödel or Blackie, this means regarding them as beings-in-becoming whose involvement in the world is necessarily grade II. We have, then, two competing conceptions of what a given continuant is essentially. In physics there is an analogous tension between those who proceed mathematically and those who proceed experimentally, or better, between those who dream of reducing physics to mathematics and those who do not. They have competing conceptions of the world.[15] Not coincidentally, the former are committed antirealists with respect to continuants and their kinds, whereas the latter are not. Modal logicians would do well to dwell on this analogy, and to wonder whether they have anything to say about *natural* kinds as such, or about the enduring things which belong to them.

I have now stated what it is to be a natural kind and defended this statement against four possible objections. That answers our first question. As for the second—whether there are in fact any natural kinds—it will be necessary to combine the more naturalistic account of continuants of chapter 4 with the more logical account of universals in chapter 3.

## VARIETIES OF NATURAL-KIND REALISM AND ANTIREALISM

Does the concept of a natural kind have any real application? I try to answer this question in two steps. First, what are the realist and antirealist options available to us? Second, which of these options should we choose? The focus throughout will be on straightforward applications of the concept to individuals, not masses, that are continuants and not merely continuant-like things, four-dimensional occurrents, processes, states, or events.

The realist and antirealist options available to us are again four in number, but they correlate only imperfectly with our options in chapter 3, where we sought to ascertain whether the concept of a universal had any real application.

Unqualified realism, *or the proposition that natural kinds exist both as classes and as types.*

Let there be continuants *a, b,* and *c*. Suppose that what each *is* exemplifies a real type $K_t$. Then each of them also belongs, in virtue of its nature, to the corresponding class $K_c$. $K_t$ is a real natural-kinds type, $K_c$ a real natural-kinds class. Hence K is real both as a type and as the class whose membership is determined by that type; it is a natural kind in the full as well as primary sense of the term. Unqualified realists claim that there are natural kinds in this sense. They are committed to the existence of universals (H4) and the existence of continuants (H5).

Do *a, b,* and *c,* as substances, exemplify just one universal? One might think not, on the grounds that what each *is* is a set of essential properties, and that it exemplifies a specific type $K_t$ only if it exemplifies some generic type as well. As noted in chapter 4, however, what a continuant *is* cannot be identified with a set of properties, not even the set of all and only its necessary properties, since what it is cannot be reduced to how it is, and since one cannot regard its essence as composite without entering the Bradley Regress.[16] Moreover, as noted in chapter 3, a determinate (species) and the determinable (genus) "in" it cannot both be entities. So if we judge an exemplified universal to be a specific type, then any more generic type it appears to exemplify must be regarded as conceptual rather than real; and if we judge this universal to be a generic type, any more specific type it appears to exemplify must be regarded as conceptual rather than real. This rules out hierarchies of natural-kind universals, but not all hierarchical orderings of natural-kinds, since the concepts employed may be well founded in reality.[17]

Should we then distinguish between substantive universals, which are instantiated as well as exemplified by what continuants are, and attributive universals, which are instantiated but not exemplified by how these same continuants are? Not necessarily. For as noted in chapter 3, there appear to be non-eidetic as well as eidetic conditions of instantiation. The distinction between substantive and attributive may be grounded not in the universals themselves but in the fact that continuants are substances with attributes, and it is the distinction between what such things are and how they are that leads unqualified realists to regard some universals as sortals, others as nonsortal.[18]

Continuants *a*, *b*, and *c*, resemble one another in many different ways. It is as substances, however, that they belong to the natural-kind $K_c$. So let us distinguish resemblances based on how they are from those based on what they are. Call the latter "s-resemblances." $K_c$ then is an s-resemblance class. Must the s-resemblances among its members be exact? If so, natural-kinds classes are probably smaller than we ordinarily suppose. But if not, if the requisite s-resemblances may be inexact, then it must be the case that tokens of $K_t$ need not instantiate it "perfectly," so it is probably the case that $K_c$ does not have sharp boundaries. May $K_c$ also be a lineage or kinship group? Yes, and if it is, this may help to explain why its members s-resemble one another. But even so, it is as an s-resemblance class and not as a genealogically related kinship group that $K_c$ is a natural kind.

Since they are composed of continuants and continuants are entities, natural-kind classes are not themselves entities.[19] Yet they are real inasmuch as their members and the "substantive" universal that determine them are real. And unlike abstract sets, these natural classes can gain or lose members without ceasing to be. What if $K_c$ were to lose all of its members? Then, apparently, it would cease to be. But what if, at some later time, there came into being a $K_c$* whose members severally instantiated the very same $K_t$ as all and only members of $K_c$ once did? Should we then say that $K_c$* and $K_c$ are diverse natural-kind classes determined by the same natural-kind type, or that a natural-kind class can exist intermittently?

If a continuant *a* belongs to the natural kind $K_c$, it belongs to $K_c$ necessarily, since the universal determining this class is the $K_t$ instantiated by *a* and every other member of the class. It is impossible, moreover, that *a* belong also to some other natural-kind class $K_c$*. For if it did it would instantiate two different "substantive" universals $K_t$ and $K_t$*, which is absurd.[20] But what then should we say about the following case? Let *a* be an atom having atomic number 55. We classify it as a Cesium atom. This, according to Brian Ellis, is its "kind-essence." It sometimes happens that by undergoing $\beta^-$ decay an atom of Cesium becomes an atom having atomic number 56.[21] We then classify it as a Barium atom. Barium atoms have a different kind-essence. Yet according to Ellis, the atom which has undergone this process remains the same continuant; it has the same "individual-essence." We should reject, however, either the supposition that *Cesium* and *Barium* are natural kinds or the supposition that $\beta^-$ decay is a nonsubstantial change for the continuant *a*. Ellis' distinction between individual- and kind-essences would then disappear, along with the possibility of natural transubstantiation.[22] One would realize that issues concerning the reality of natural-kind classes cannot be treated apart from issues concerning the ways in which enduring things do or do not remain the same as they change.[23] Again, as I have tried to show in these chapters, the metaphysics of natural kinds depends partly on the metaphysics of individual continuants.

Continuant *a* and the universal $K_t$ it instantiates are both entities. As noted in chapters 3 and 4, this cannot be the case if universals are immanent in their instances; for then $K_t$ would the nature common to *a*, *b*, and *c*, hence these three entities would be one entity.[24] Therefore, with respect to "substantive" universals at least, unqualified realists must accept a one-over-many theory. Yet our foremost natural-kinds realists, Aristotle and Mill among them, have steered clear of such "Platonic" implications. Aristotle in particular seems to have been a one-in-many theorist who made continuants the genuine entities. As many scholars have noted, however, his "substantive" universals (species) seem to exist only in thought as concepts well founded in the natures of the continuants to which they apply.[25] Our foremost natural-kinds theorists appear not to have affirmed the existence of natural kinds in the full sense of the term. They do not appear to have been unqualified realists.

Qualified realism, *or the proposition that natural kinds exist as classes but not as types.*

Let there be continuants *a*, *b*, and *c*. There is no real type which they exemplify as substances. Yet suppose they s-resemble one another. Then *a*, *b*, and *c* belong, in virtue of their natures, to the class K. This s-resemblance class is a natural kind in the primary but not full sense of the term. Qualified realists affirm the reality of natural kinds in this sense. Their position commits them to the existence of continuants (H5) but not to the existence of universals (H4). They may be resemblance theorists across the board, or else one-in-many or one-over-many theorists who deny the existence of "substantive" universals.

Although there is no real type $K_t$, it may nevertheless be possible for us to form a sortal concept by means of which to decide whether some given continuant *x* does or does not belong to K. It may even be possible to order all natural kinds into a single taxonomy by means of nested classificatory principles of this sort.

Though real, K is not an entity since it depends essentially on some, though not all of the individuals that are its members. These entities may constitute a lineage or kinship group whose members tend to breed true; but again, it is as s-resembling particulars that they constitute the natural kind K. And again, unlike an abstract set, it may gain or lose members without ceasing to be. Imagine, however, the following scenario. Continuants *a*, *b*, and *c*, the only members of K, all cease to be. Then, at some later date, continuants $a^*$ and $b^*$ come into being and constitute, in virtue of their natures, the natural kind K*. Suppose $a^*$ and $b^*$ s-resemble *a* or *b* or *c* as closely as they s-resemble each other. Should we then count K and K* as two natural kinds, or as one intermittently existing natural kind, even though there is no real type $K_t$ that its members severally instantiate?

Are *a*, *b*, and *c* members of a single natural kind K only if they s-resemble one another exactly? If so, then not only are natural kinds probably much smaller than we think, in the absence of any real type $K_t$ we have reason to expect this condition will rarely be met. But if *a*, *b*, and *c* belong to K even though they s-resemble one another only inexactly, we have little reason to expect that natural kinds will be disjointed, sharply distinguishable from one another. One way out of this difficulty would be to say that all continuant individuals constituent a single natural kind. Another would be to say that there are as many natural kinds as there are individual continuants. But leaping to either of these extremes is too easy; it would be tantamount to evading the difficulty. A more promising way out is to observe that quarks and leptons may resemble one another so closely that one cannot tell them apart. Let us make these the members of exemplary natural kinds. Then the boundaries separating one kind from another are quite sharp. But quarks and leptons are not exemplary continuants. The exemplary continuants are animals and plants, but these rarely resemble each other so closely that we cannot tell them apart and the natural kinds they appear to form have boundaries that are far from sharp. We thus confront a dilemma: there are exemplary continuants without exemplary natural kinds or exemplary natural kinds without exemplary continuants. I shall return to this difficulty in chapters 6 and 7.

Qualified antirealism, *or the proposition that natural kinds exist as types but not as classes.*

Suppose there is a real natural-kind type $K_t$ and yet no real class $K_c$, because continuants are no more possible in the actual world than are Euclidean cubes, or because there are in fact no continuants in the actual world, or because existing continuants s-resemble one another so inexactly that any division of them into discrete classes would be arbitrary. Qualified antirealists are committed to the existence of universals (H4) but not to the existence of continuants (H5). And if they do admit continuants as well as "substantive" universals, and refuse to say that they constitute a single natural kind, or that there are as many natural kinds as there are continuants, they may then argue that one can do no more than note the s-similarities and s-dissimilarities among them.

Why do I call this position "qualified *anti*realism"? For three reasons. First, it is directly opposed to qualified realism. Second, although natural kinds may be understood both as types and as classes, it is as classes that they are first of all understood in the Millian tradition, the Aristotelian tradition, and in our everyday conception of them.[26] Thus natural kinds without natural-kind types are a lot easier to swallow than natural kinds without natural-kind classes. Third, it is as classes that natural kinds are of this spatiotemporal world. Universals are nonspatiotemporal, transnatural. Therefore, to claim the $K_t$ is a real type that determines no real class $K_c$ is to give up our only reason for regarding $K_t$ as a *natural*-kind type. Such types are better understood as permanent possibilities of natural kinds.[27] And according to qualified *anti*realists, these possibilities are never actualized.

Unqualified antirealism, *or the proposition that natural kinds exist neither as classes nor as types.*

Those who accept this proposition are free to affirm or deny the existence of universals (H4) and the existence of continuants (H5). If they reject H5, it follows necessarily that there are no natural kinds in the primary sense of the term. If they accept H5 and deny H4, they must deny that there is any basis in continuants themselves for whatever substantive sortal concepts we apply to them; all so-called natural-kind classes are really conventional or stipulated classes projected into the spatiotemporal world. And if they affirm H4 together with H5, they must deny that there are any "substantive" universals; all so-called natural-kind types are really sortal concepts projected onto eidetic space.

These, in sum, are the four options concerning the ontic status of natural kinds: qualified and unqualified realism, unqualified and qualified antirealism. There are no other options. Our enumeration is complete.

## A SIXTH HYPOTHESIS?

All four options are compatible with H4 and H5. All four accord with the phenomena. So far, then, we have no reason to favor any one of them over the others. Yet just one of them is correct, we suppose, since there are no other competing options. Recall, however, the discussion of realism and antirealism in chapter 2. There we saw how protometaphysical antirealists can deny that there is just one way the world is. What if, contrary to hypotheses H1, these antirealists are right? Then we should not suppose that just one of these four metaphysical options is absolutely correct. But rejecting H1 at this point, in order to avoid the agony of choice, would be cowardly. Better to proceed on the supposition that protometaphysical realism is correct, and see where it gets us.

Although qualified antirealism may be correct, accepting it seems hardest to justify, even if we do have good reason to accept H5. For according to this option, continuants falling within our purview either do not exemplify universal archetypes at all or else exemplify them so "imperfectly" that there are no natural-kind classes. Either way, we have no sure way of moving cognitively from things of this world to those transworldly universals, hence no obvious way of discovering that they even exist.[28] Although the "Platonism" in qualified antirealism is not necessarily out of place, the epistemological difficulties it brings are generally much greater in natural philosophy than in mathematics or ethics. For us, then, the main options right now are realism, whether qualified or not, and unqualified antirealism. Let us rewrite them in our usual manner.

> H6. There are natural-kind classes, and maybe natural-kind types, in the primary sense of the term.
> H6′. There are no natural-kind classes or natural-kind types in the primary sense of the term.

The question before us is whether one should accept H6 (natural-kinds realism) or H6′ (natural-kinds antirealism).

We have several reasons for favoring H6. First, common sense strongly inclines us to accept it. As noted in chapters 1 and 2, the things we experience come already typified, among them things we take to be natural, and these seem for the most part to fall naturally into kinds, without any forcing on our part. Natural-kinds realism is implicit in the folk metaphysics of everyday life.[29] Second, as noted in chapters 3 and 4, we evidently have some ability to recognize universals in particulars, and some ability to identify and track continuants through their manifold variations.[30] These abilities appear to be partly innate. Also apparently innate, as Quine noted, is our ability to allocate things to different places in "similarity space."[31] What then prevents us from finely attuning ourselves to the substantial similarities and dissimilarities among continuants we recognize, and of classifying them accordingly? And may we not find that these disjoint s-resemblance classes are determined by the diverse "substantive" types their members exemplify?

Third, there is remarkable cross-cultural agreement concerning which plants and animals belong to which kinds. It is hard to see how these kinds could be merely conventional. Also remarkable is the ability of some bird watchers to identify, often at a glance, rare specimens they see in the field. How is their expertise to be explained? According to Jeremy Mynott, their discernment of species is the result of "active attention," "informed expectation," and "ambition of imagination," all operating in balanced combination. The first and second ingredients come with interest and experience, as do habits. The third, which distinguishes the great from merely good discriminators, is what takes one beyond all preconceptions about the specimens at hand to the specimens as they are in reality.[32] To put Mynott's thesis in an idiom more familiar to us: We identify continuant-like things perceptually and sortally, and then, using the evidence of each these cognitive resources to refine and adjust that of the other, we zero in epistemically on both the real continuants and the natural kinds to which they belong. While zeroing in, our discoveries or determinations are perceptually and sortally dependent and yet, as noted in chapter 4, perceptual or sortal dependence does not entail perceptual or sortal relativity.[33] We can in principle transcend prior experience and typological prejudice and come to understand things natural as they really are. Sometimes, apparently, some of us do so.

These are all reasons for accepting H6. But there are also reasons against. First, as noted in chapters 1 and 4, the natural-kinds realism implicit in our "natural attitude" is not so secure that encountering hard cases will rarely, if ever, lead one to question it. And once our confidence in it has been shaken, we realize that our everyday understanding of the world muddles crucial distinctions between continuants and continuant-like things, between what and how continuants are, between substantive and nonsubstantive sortals,[34] and between natural-kind classes and natural-kind types. What exactly is it, then, that we are tracking when in our rough and ready ways we seem to be tracking individual continuants? And what exactly are the objects we ordinarily take to be their natural kinds? These questions are disconcertingly difficult to answer.[35]

Second, we have encountered three waves of difficulty—regarding first the existence of universals, then the existence of continuants, and now the existence of natural-kind classes or types. The first wave did not keep us from tentatively accepting H4. The second wave, though greater, did not keep us from very tentatively accepting H5. Now we confront a third wave of difficulty: it is far from clear that the similarities and dissimilarities among existing continuants are such that these entities form disjoint s-resemblance classes, or that they exemplify diverse universals in virtue of their natures. And supposing they do, it is far from clear that we have the time and ability to discern either these classes or those types. These waves are cumulative. Even if we reject realism with respect to natural-kind types, the third wave still piggybacks on the second, whose force was already great. It seems that our acceptance of H6 would have to be so tentative as to be no acceptance at all.

Third, when we look to the history of natural philosophy for guidance, we find that natural-kinds antirealism has been the predominant view. And again, when we look to fundamental physics we find reference to elementary particles that are not exemplary continuants and probably not continuants at all, since we cannot single out and track them. And in biological science, as distinguished from natural history, we find reference to species which may seem at first to be natural-kind classes but in fact are not, since common ancestry rather than any resemblance is what holds them together. Indeed, as chapters 6–7 will show, it is by no means obvious than the concept of a natural kind (in the primary sense of term) has any well-established theoretical use in the natural sciences today. Our everyday acceptance of H6 could still be justified on the grounds that it has considerable survival value—which would also help to explain why different societies so often agree about which natural things belong to which natural kinds—but then the kinds we take to be natural in everyday life are really pragmatic kinds, Baconian idols which are helpful if not absolutely indispensable for the perpetuation of *Homo sapiens*, the specific kinship group to which we belong and which we tend naturally to confuse with the specific resemblance class we call "humankind."[36]

These are all reasons for rejecting natural-kinds realism (H6). Are they reasons enough for accepting natural-kinds antirealism (H6′)? I think not, for they do not manifestly outweigh the reasons for accepting H6 and rejecting H6′. It remains to be seen just how strong the last-mentioned reason is: the penchant for natural-kinds antirealism that grips the biological and physical sciences today may prove questionable, and a minority view in natural philosophy may prove correct. So, in the light of all these pros and cons, which of these two competing hypotheses should we accept? Common sense alone, it appears, is not up to the task of answering this question.

## JUNCTURE

At this point an ancient skeptic might urge us to suspend judgment, thus leaving unanswered the question whether there are any natural kinds. Yet let us not be hasty. Recall that natural philosophy today can take guidance from three sources: the folk metaphysics of common sense, scientific theories, and philosophical traditions. Of these, the first is always the primary resource for us, and the one most befitting our initial investigation. It was under the gentle guidance of this resource that we first wondered which kinds are in fact natural, and whether there are any natural kinds; and having raised these questions we realized that we needed also to find out what it is to be natural kind. As philosophers, moreover, we were aware of the contemporary renaissance in natural philosophy. Our aim accordingly has been two-fold: to give an account of natural kinds by using the resources available to us, and to show how natural philosophy today can be rejuvenated (chapter 1).

Implicit in our robust sense of nature are three assumptions we accepted as working hypotheses: that what there is does not depend on how we think or speak about it (H1), that the natural world is real (H2), and that our everyday understanding of things natural is heuristically if not cognitively reliable (H3). So inspired, we set sail in pursuit of our two-fold aim. Immediately, however, we were blown off track by a pair of strong philosophical objections, one emanating from a variety of antirealism, the other from a variety of naturalism. It took us a while to get back on track (chapter 2).

Natural kinds are to be understood as classes and as types. Understanding them as types led us first to argue for a variety of realism with respect to universals (H4), and then to determine which of these real types, if any, might have as tokens those entities which could be members of natural-kind classes. We then realized that only by making much ado about nature could we avoid splintering the concept of a natural kind into eidetic and non-eidetic bits. This led us to investigate continuants as intrinsically mutable entities deeply involved in the ever-changing material world. We realized that only by making much ado about continuants as beings-in-becoming could we avoid turning them into intrinsically mutable 'dividuals' or into individuals only incidentally mutable. The ensuing ontological and epistemological difficulties were formidable, yet we tentatively accepted as another working hypothesis (H5) the widespread assumption that continuants exist (chapters 3 and 4).

We were then in a position to articulate the concept of a natural kind and make clear its distinctive role in our thought. Having thus answered one of our two leading questions, we tried to answer the other—"Are there in fact any natural kinds?"—by setting out the realist and antirealist options available to us and selecting one of them. The result was indecision (chapter 5).[37]

Two possible courses now lie before us. One is to keep rowing under the guidance of pre-scientific naturalism, in the hope that something as yet unnoticed will turn up, which will enable us to move beyond our present impasse. The other is to place ourselves under the guidance of scientific theories, in the hope that inspiration from these quarters will give us a second wind. The first way would keep us close to what motivated our inquiry at the outset, making it less likely that we will commit fallacies of irrelevance. But that danger appears much diminished now, as does the capacity of common sense to keep moving us toward our goal. I propose that we go wide by turning to our second major resource for guidance. Seeking inspiration from natural science seems particularly apt at this stage, since we tend to regard it as authoritative in empirical inquiries of this sort, and since pre-scientific naturalism (the "manifest image") and scientific naturalism (the "scientific image") form for us a contrastive pair. In the next two chapters I intend to show that scientific theory, properly used, does indeed provide an affirmative answer to our second leading question: there are natural kinds in the primary sense of the term, but they are generally not among the kinds acknowledged in the folk metaphysics of common sense.[38]

## NOTES

1. Call all the gold in the world G1. Suppose G1 perishes, and then, at some later date, some gold (G2) comes into being. If G1- and G2 were individuals, it would be reasonable to ask whether they are one continuant or two: recall the case of Lazarus and Lazarus* in chapter 4. But since gold is presumably a mass, G1 and G2 are nothing more than spatiotemporally scattered portions of one and the same continuant.
2. Laycock 2006; and Zimmerman 1995, 53ff.
3. See chapter 1 of this volume.
4. Mill 1973, I, 27–40, 46f, 133 and context.
5. Venn 1923, 55ff.
6. The principal difference between the two traditions is this: the Aristotelians were very interested in the ontic status of natural-kind types, and in continuants as entities, whereas the Millians generally speaking are not. In this one respect, if no other, my account is more Aristotelian than Millian.
7. Hacking 2007, 202–39.
8. Cf. Magnus 2012, 68–83.
9. Cf. G. E. L. Owen, "Logic and Metaphysics in Some Earlier Works of Aristotle," in *Articles on Aristotle 3: Metaphysics*, ed. J. Barnes, M. Schofield, and R. Sorabji (London: Duckworth, 1975), 13–32, M. Ferejohn, "Focal Meaning and the Unity of Science," *Phronesis* 25 (1980): 117–28, and G. Lakoff 1987, 91–114, 153f, 204f, 287–91.
10. Cf., e.g., de Sousa 1984, 565–80 with Dupré 1981, 82 and context; 1993, 22, 60–84, 103–05; and 2002, 29–49. See also Boyd 1991, 127–48, and 2010, 212–34; and Magnus 2012, 43–49. That the plausibility of Dupré's pluralism depends partly on the fact that he takes his cue from biology, see: M. Slater 2005, 24 (cf. 31, 39). That the sciences themselves, in the absence of any essentialism, may serve to constrain this proliferation of uses, see: Khalidi 2015, 42ff.
11. E.g., Plantinga 1974, 55–87; Kripke 1980, 114n57; Wiggins 1980, 215f; Brody 1980, 116–23; Almog 1991, 228; Mackie 2006, 156f.
12. Chapter 3 of this volume.
13. Or else one feels that no line is to be drawn here, in which case individual essentialism gives way to haecceitism. See P. Mackie 2006, 150–68.
14. Cf. Harré and Madden 1975, 14f, 19–21, 27f, 47, 53, 97, 118–39; Ellis 2001, 53f, 172f, 220f, 231–37.
15. There are analogous tensions in the philosophy of mind and in moral and political philosophy. For instance, some like to ask what would happen if your brain was transplanted into my body or to study the Ticking Time-Bomb scenario, whereas others prefer to keep the focus on real people in real-life situations.
16. Chapter 4 of this volume.
17. Chapter 3 of this volume.
18. Chapter 3 of this volume.
19. Borgesson 1999, 873–79.
20. Cf. Wiggins 1980, 115f; Ellis 2001, 97–101.
21. In $\beta^-$ decay the weak interaction converts a neutron into a proton while emitting an electron and an antineutrino: $n \rightarrow p+ + e- + \bar{\upsilon}_e$. When this occurs within the nucleus of an atom of Cesium$^{137}$ it becomes an atom of Barium$^{137}$.
22. Cf. Ellis 2001, 11, 92, 237–41, 245–47. Compare Bigelow 1999, 56–76; and L. Clapp, review of Ellis' *Scientific Essentialism* in *The Philosophical Review* 111 (2002): 593f.
23. Compare Wilkerson 1995, 167.
24. See chapters 3 and 4 of this volume.
25. Those in the Millian tradition of natural kinds have generally said very little about such kinds as real types. A notable exception: H. H. Price 1953, 7 and context. But Price, like Aristotle, is no "Platonist" in his conception of universals.
26. Putnam's extensionalist construal of natural kinds accords with this view; consider, e.g., "The Meaning of 'Meaning,'" in Putnam 1975, 234. Thus, for individual continuants vis-à-vis natural kinds, the membership relation takes precedence over

the instantiation relation—even though, according to unqualified realists, specimens are at once members and instances; and even though, according to some qualified realists, we would not take specimens to be members if we did not also take them to be things to which a substantive sortal concept applies.

27. Cf. Newman 1992, 41–69, esp. 50.

28. See chapter 3 of this volume.

29. See chapters 1 and 2 of this volume.

30. See chapters 3 and 4 of this volume.

31. Quine 1969, 122f.

32. Mynott 2009, 54ff, 109ff.

33. See chapter 4 of this volume, and Wiggins 1980, 87.

34. On the importance of this distinction, cf. Wiggins 1980, 27, 64. On the difficulty of distinguishing substantive from phase sorts in particular, cf. Darwin's work on cirripedes and N. Knowlton, "Sibling Species in the Sea," *Annual Review of Ecology and Systematics* 24 (1993): 189–216.

35. One finds a similar murkiness even in some philosophical accounts of the issue. For instance, in Kripke 1980, when he extends his direct-reference theory of proper nouns to "natural kind" terms, he fails to make clear just what it is that these common nouns designate (rigidly?). And Wiggins appears sometimes to be a resemblance theorist, at other times an immanent realist with respect to universals; cf. his 1980, 10, 47, 83, 85, 88, and 72, 120, 139, 169, 211 with 78n2, 129. Putnam, too, seems to have vacillated between these two positions before becoming a protometaphysical antirealist; cf. Putnam 1979, 141–42, 234–35, 249–51, 277.

36. Bacon, *Novum Organum,* I.39, 41, 45–52. On pragmatic as distinguished from real individuals or kinds, see chapter 4 of this volume.

37. Chapters 3–5 exemplify what Daniel Dennett calls "sophisticated naïve naturalism": Dennett 2013, 96–107. But I do not share his strong partiality to what Sellars has called "the scientific image," and am less interested than he in constructing a "Mercator of the Manifest Image."

38. I make use of relevant philosophical traditions, our third major resource, in a forthcoming companion volume entitled *The Aristotelian Tradition of Natural Kinds and Its Demise.*

# SIX

# Physical Kinds

It was J. S. Mill who instituted the modern tradition of natural-kinds realism. A natural kind, he said, is one whose members have inexhaustibly many common properties and these properties are connected neither logically nor by means of any ascertainable law.[1] This agrees with our sense that tigers, water, and the like are naturally property-rich, but not with the presumption that science can ultimately explain everything natural. C. S. Peirce noted that scientists try to explain natural kinds in much the same way that Mill explained man-made classes, which is to say, in terms of one or several common properties connected by a law of nature. He revised Mill's account accordingly.[2] This agrees with the presumption that science can ultimately explain everything natural, but not necessarily with the proposition (for which I argued in chapters 4 and 5) that the members of exemplary natural kinds are enduring things having natures or essences. Saul Kripke and Brian Ellis, among others, have revised Peirce's account accordingly. They agree that the members of a natural kind are material things having essences, and that terms designating natural-kind terms are to be defined scientifically, yet they differ in two fundamental respects. First, Kripke's account is primarily semantic (of natural-kind terms), whereas Ellis' is primarily metaphysical (of what such terms designate). Second, Kripke draws his natural-kind terms (e.g., "water," "tiger") mainly from prescientific discourse, whereas Ellis draws his (e.g., "cesium," "electron") directly from scientific theory.

Central to Kripke's account is the dichotomy that often obtains between those who name a natural kind without yet knowing what it is (namely, laymen) and those who can tell us what it is (namely, scientists). Ellis, however, regards any such dichotomy as peripheral at best. To mark this difference, I take Ellis' account of natural kinds to represent *scientific* essentialism, Kripke's account to represent *prescientific* essentialism. In chapter 1, I drew attention to essentialism of both sorts, but then set it aside on the ground that we were not yet ready to examine the scientific theories to which they appeal. Now we are. I begin by considering how two prescientific essentialists, Kripke and Putnam, understand the link between the world of everyday discourse and the world of scientific theory.

## STEREOTYPES AND STEREOTOKENS

To see how Putnam and Kripke link everyday discourse with scientific theory, consider their accounts of the term "water." Ordinarily we use this term to refer to the stuff in this glass, the stuff raining on us, and so on. The stuff in each case is water, we say, and if asked what water is, we would probably say that it is a transparent, colorless, odorless liquid. Kripke and Putnam distinguish strictly between that which "water" denotes, water, and the everyday connotations associated with this term.[3] Among these connotations is the description just now offered, which we use daily to identify samples of water. Rough and ready criteria of this sort Putnam calls "stereotypes."[4] Though reliable enough for most practical purposes, these stereotypes sometimes fail us. Already we agree that not everything water-like is water, and that ice cubes and steam are water even though they are not especially watery, and that being a transparent, colorless, liquid is not what water *is*. We may then be inclined to think that "water" is a family-resemblance term denoting an open-ended class of natural kinds determined by an open-ended stereotype. According to Kripke and Putnam, however, the stuff denoted by "water" is determined instead by that to which the name "water" was originally applied, and by anything else just like that thing in its nature or essence. Hence the denotation of the term remains invariant, fixed, or rigid under connotative variation.[5]

Discovering the nature of water and other natural kinds is a job for scientists. They seem to have been remarkably successful. Take for example the statement, "Water is $H_2O$." According to Kripke, this is an identity statement whose flanking terms are names designating rigidly one and the same substance. Hence, if true, as apparently it is, this statement is necessarily true: nothing could be water and not be $H_2O$. This *de re* necessity was discovered empirically. That water is $H_2O$ is therefore an *a posteriori* necessary truth.[6] In this case, as in others, we witness what Putnam calls "a division of linguistic labor": pragmatically-oriented laymen have named a natural kind (water), whose underlying nature theoretically oriented scientists have investigated and apparently discovered. The term "$H_2O$," whose meaning is fixed, has in theory supplanted the stereotypes we ordinarily associate with the name "water."[7]

There are difficulties with this proposal. I mention six. First, how can one attain knowledge of *de re* necessities by empirical means? Epistemologists have not shown this to be impossible, yet how it is possible remains to be shown.[8] Second, is it true that the distinction between denotation and connotation mirrors the protometaphysical realist's distinction between ontic and semantic? We shall encounter cases where the connotation-shift has been so drastic that the reference itself seems to have changed, even though the reality itself has not. Third, is Putnam's account of the division between colloquial naming and theoretical determination acceptable? Keith Donnellan and others have argued that it is simplistic.[9] But these three difficulties are peripheral to my present inquiry. The next three are more serious in the context of our present inquiry.

Fourth, "Water is $H_2O$" is not a statement of numerical identity, as Kripke avers, but an essence statement. And unlike "This stuff is water," it appears to be metaphysical rather than protometaphysical. I tentatively classify it as a nature statement: it declares, or purports to declare, what the stuff we call "water" is in its nature.[10]

Fifth, "$H_2O$" is not a name but a description, a connotation, one which these new essentialists take to be rigid: it seems to describe what water *is*.[11] Such statements used to be called "real definitions." That man is a rational animal remains a favorite textbook example. The connotation "rational animal" supplanted, in Aristotelian natural philosophy, the nonrigid stereotypes or conventional ("nominal") definitions usually associated with the natural-kind term "man." In early modern natural philosophy or physics, such definitions fell by the wayside. But then they returned, unheralded, in modern chemistry and atomic theory. The new essentialists have finally recognized such statements for what they are. That water is essentially (di)hydrogen oxide, or $H_2O$, is one of their favorite examples.[12]

Sixth, Kripke and Putnam view as necessarily true the nature statements they find in scientific theory. Mill held the opposite view. In all such statements, he implied, some Putnamesque stereotype has been made authoritative in a vain effort to make nature theoretically manageable.[13] Who is right? In this chapter and the next I will give reasons for thinking that Mill is wrong, but also argue that the new essentialists, scientific as well as prescientific, have been much too quick regard as genuine many of the putative nature statements they seem to have found in scientific theory. In other words, there are natural kinds in what I have identified as the primary sense of the term, and yet many of the kinds named and described in scientific theory are not natural kinds in the primary sense of the term.

To facilitate the presentation of my argument, I make two preliminary decisions. First, the scientific descriptions of putative natural kinds as types I call "(scientific) stereotypes"; and the putative continuants that are their tokens or members I call "(scientific) stereotokens." Our task will be to identify these stereotypes and stereotokens, and to discover whether they do in fact determine natural kinds. Second, I distinguish between physical and biological sciences. The former include physical chemistry and particle physics, which I consider in the rest of this chapter; the latter include evolutionary theory and molecular genetics, which I consider in chapter 7. In the final section of chapter 7 I will assemble my reasons for thinking that *some* physical kinds, and *perhaps* even a few biological kinds, are natural kinds in the primary sense of the term.

One more thing before we begin. In their desire to ground natural-kinds realism in the physical and biological sciences, essentialists today tend to overestimate the degree to which contemporary science supports any neat typology of matter, and to underestimate the tension between such neat typologies as they do find and continuant realism, or between their essentialistic view of natural *things* and the nomocentric view of the natural *world*. To show all this, while also showing that there are natural kinds in the primary sense of the term, I must survey a range of scientific theories. Those interested in metaphysics of natural kinds, but unfamiliar with some of these theories, will find what follows hard going in places. For them I have tried to keep technical matters to a minimum. Those well acquainted with one or another of these theories will find what follows to be rather superficial in places. I ask them to bear in mind my major aim, which is to show how scientific theory does and does not support an affirmative answer to the question whether there are any natural kinds in the primary sense of the term.

## FROM SUBSTANCES TO PARTICLES

"Chemistry is the science of substances—their structure, their properties, and the reactions that change them into other substances."[14] Substances are of two sorts: elements and their compounds. All are regarded as continuants belonging to disjoint kinds, which theoretical chemistry defines in terms of microconstitutive or microstructural types. These are among the principal stereotokens and stereotypes in chemical theory. Are they also natural kinds? We must look more closely, to make sure we have not hit fool's pay-dirt. I begin with water.

Although chemists agree that water is $H_2O$, they do so in full awareness of such complications as the following. (1) Liquid water contains the polymeric forms $(H_2O)_2$, $(H_2O)_3$, and so forth. Would it not be more accurate, then, to say that water is $(H_2O)_n$ where $n \geq 1$? (2) Linguistically-oriented philosophers want to understand why salt water and runoff are called "water," whereas tea and human bodies are not.[15] These issues lie beyond the purview of chemistry, and yet some purity issues are germane. For instance, chemists have discovered that liquid water *cannot* be just $(H_2O)_n$, since it is "amphiprotic"; it acts as both an acid and a base, and thus reacts with itself. There is always present in it a dynamic equilibrium described by the reaction equation

$$2H_2O \leftrightarrow H_3O^+ + OH^-$$

where $H_3O^+$ and $OH^-$ are hydronium and hydroxide ions, respectively.[16] Hence, if on Putnam's Twin Earth we should find a liquid that was pure $X_2Y$, it would not be water. (3) There are three isotopes of hydrogen—$^1H$ (protium), $^2H$ (deuterium), and $^3H$ (tritium)—and more than a dozen isotopes of oxygen, three of which occur naturally—$^{16}O$, $^{17}O$, and $^{18}O$. Hence, in the samples of water we encounter, there are some eighteen specific kinds of $^mH_2{}^nO$. Should we not then say that "water" denotes a mixture of different material substances, each of which is one of a kind? (4) "Water" is a mass-noun, as are "iron," "ethanol," etc. We take these terms to denote homogeneous masses of one kind or another. Chemists, however, regard every such mass as an ensemble of particles. Hence "water" is really like "furniture," or "seed" used as a bulk-term, and not like "substance" as we ordinarily construe it. Or else it has become ambiguous, denoting one sort of stuff in everyday life, a very different sort in chemical theory.[17]

We may still say that chemistry is a science of substances.[18] We may also say that it describes the micro-constitutive and micro-structural *natures* of these substances, in terms of which their bulk properties are to be explained: why, for instance, water disintegrates above 500° C, why liquid water is a solvent, why it is less dense as a solid than as a liquid, why there is glassy as well as crystalline ice, and so forth.[19] But then we are no longer using the word "nature" as I did in chapter 4. For any portion of water is an ensemble of similar particles interacting with one another in similar ways. These ensembles are compositionally grade III. And although such objects may be continuant-like, as water and other chemical substances certainly are, they are not genuine continuants; they do not have natures in virtue of which they remain invariant under natural or broadly logical variation. Hence there is no such thing as the *essence* of water, or of iron, carbon, or any other chemical substance. Scientific essentialists must look elsewhere for the stereotokens and stereotypes they seek.

Had we started with mankind, regarded as a scattered population of humans and concluded that no such ensemble is a genuine continuant, we would then have focused on those individual humans. Having in fact started with water and come to regard it as an ensemble of particles, it seems reasonable to focus now on these particles.

An $H_2O$ molecule is not just two hydrogen atoms plus one oxygen atom. Nor are two molecules necessarily the same in kind if their constituents are the same in kind. For example, "$C_2H_4O$" is the formula for vinyl alcohol (ethenol) and acetaldehyde, two distinct molecular kinds; and

"$C_2H_6O$" is the formula for ethyl alcohol (ethanol) and dimethyl ether, also different in kind. What these formulae hide is the intrinsic structure of the molecule and it is in virtue of this structure, apparently, that a molecule is one thing and not just several things in close proximity.[20] Two molecules may be very similar in structure yet differ in kind. $H_2S$ (hydrogen sulfide) and $H_2O$ (hydrogen oxide) have the same electronic structure and their bond-lengths and bond-angles are not very different, yet they differ in kind and have very different bulk properties.[21] $H_2O$ and $D_2O$ may also be regarded as different in kind, even though their bond-lengths and bond-angles differ only slightly. And the same is true of ortho- and para-$H_2O$, where the only structural difference is in the spin-orientations of the constituent hydrogen nuclei; and of $H_2O$ and anti-$H_2O$, where there is no structural difference at all. Hence molecules belong to the same specific kind just in case (1) they are what chemists call "isomers," i.e., they have the same constituents, and (2) they are what we may call "isomorphs," i.e., they have the same intrinsic structure. These molecules also differ in number. Hence they are tokens of the type $H_2O$ just in case (1) each has three constituents, two of type H and one of type O, and (2) each has an intrinsic structure of type S. Together these conditions describe generally what each of these molecules is in its nature. And what they describe is apparently a *de re* necessity.

Chemists do not talk this way, but they do require that molecules belonging to the same specific kind be isomers and isomorphs, and what they call "structure" is central in their explanation of a free molecule's properties and behavior. To us their conception seems hylomorphic; we could reframe it in Aristotelian terms, taking the constituents of a molecule to be its material cause, the intrinsic structure to be its formal cause. Small wonder, then, that "molecular structure," like animal "form," is so hard to understand.

Of the various ways in which chemists represent it, most are pictorial: Lewis diagrams, ball-and-stick configurations, etc. These representations are very useful but not very accurate. For greater accuracy, chemists resort to quantum mechanics. The structure or global "state" of a molecule, that in virtue of which it is one complex thing and not several diverse things, can be described quantum-mechanically by means of a time-independent wavefunction. To help us become acquainted with wavefunctions, educators use any number of visual aids. In the end, however, what such functions represent is not visualizable: molecules have no determinate shapes or sizes; there are no bonds in addition to the things bonded, and consequently no real bond-lengths or bond-angles; and its constituents do not move about one another in determinate orbits.[22]

Thus, in theory, for any stereotoken of specific molecular kind K, its corresponding stereotype is some specific wavefunction $\Psi_K$. In practice, however, the precise $\Psi_K$ is rarely specifiable. What chemists usually do is start from a simplified model of the molecule in question, regard its nuclei as fixed even though they are not, and then, using a specification of the Schrödinger wave equation, they compute the molecule's electronic configuration (their main concern as chemists) by proceeding in a stepwise fashion familiar to students of perturbation theory.[23] The result is an approximation to the real $\Psi_K$. Such proximal wavefunctions are our most accurate stereotypes of specific molecular kinds or their members—the best theoretical descriptions of what such things are in their natures.

What assurance do we have that the corresponding stereotokens are genuine continuants? From chemists we learn the following. First, a free molecule abides in a state of dynamic equilibrium; it fluctuates all the time, necessarily, and yet its mathematizable structure, its nonvisualizable *form*, remains unchanged as long as it exists.[24] Second, the electrons belonging to it have been "delocalized"; they no longer belong to the atoms that have become its constituents, and as constituents they are no longer entities in their own right, having been virtualized, potentialized as it were, in the molecule to which they now belong.[25] Third, in virtue of its structure, a molecule exemplifies mathematical symmetries and causal connections that set it apart from its surroundings.[26] Hence metabolically, compositionally, and situationally its involvement in the world appears to be grade II rather than grade III. In other words, it appears to be a genuine continuant.

There are difficulties, of course. For what are we to make of the fact that the constituents of its constituents cannot be tracked, so that when two molecules come together, there is no way of telling whether they have exchanged electrons? And does the unity of an $H_2O$ molecule really differ in kind, not merely in degree, from that of an ice cube?[27] As noted in chapter 4, however, similar questions arise for any continuant-like object, including such everyday things as Blackie the squirrel. If we have reason to believe that continuum realism is true, and we do, I see no reason in principle for refusing to classify free $H_2O$ molecules as individual continuants. Nor do I see any reason for denying that $H_2O$ molecules constitute a natural kind in the primary sense of the term, if indeed they are essentially isomorphic as well as isomeric. That chemists are not particularly interested in such kinds, or in their members as situationally independent individuals, is neither here nor there.

Can we say the same of atomic constituents? While it is difficult in practice, at least, to identify and track free atoms, the stereotypes corresponding to these stereotokens are remarkably clear and distinct.[28] Consider a periodic table of elements with its neat array of disjoint places, one for each elemental kind, to each of which a positive integer is assigned. As a cardinal, this number it tells us how many nuclear protons the element has, while as an ordinal, it tells us where the element is in the ordered list of possible elemental types. For every such place there is one such type and for every such type there is one such place. The periodic table is a natural-kinds realist's dream come true, or so it seems. Yet when we look more closely, questions arise.

First, are these elemental types genuine universals? The answer appears to be "yes." But what of those places in the table that are empty in the sense that no tokens of that type have been discovered, or have occurred, or in the sense that no such tokens will ever occur, because their nuclei must have too many protons, or for some other reason?[29] Only the last-mentioned possibility would require an *ante res* theory of universals, which few scientific realists are ready to accept. Those who reject it may still regard the remaining elemental types as real. Or they may regard all such types as concepts well founded in the relevant resemblance classes. These classes are disjoint in any case, since nuclear charge is quantized. The presence of well-defined but unfilled places in the periodic table may be regarded heuristically as a stimulus to further research and not metaphysically as an affirmation of unexemplified universals in eidetic space.

Another question concerns the fact that some elements have been discovered by synthesizing them in the laboratory. Are these, too, natural kinds? The answer appears to be "yes" if the elemental stereotokens are continuants. For then, regardless of how they came to be, they belong to discrete resemblance classes in virtue of their chemical natures or essences.[30]

Another, deeper question concerns these chemical essences. To set it up, we note three apparent anomalies in the periodic table. First, there is the potentiality problem. Elemental hydrogen, regarded as a class, includes not only H atoms but $H^+$ and $H^-$ ions as well as those bound molecular constituents to which we assign the letter "H."[31] Consider a free protium$^+$ ion. Call it "*h*." Regarded physically, *h* instantiates *the proton*; it is a subatomic particle, complete in itself. Regarded chemically, *h* instantiates *the hydron*[32]; it is potentially but no longer or not yet actually a hydrogen atom, since it lacks an electron that would make it complete. Which is *h* as it is in reality? And why should we regard $^1H^+$ ions and $^1H$ atoms as being exactly alike in their natures when in fact they are neither isomeric nor isomorphic with each other?

Second, there is the isotope problem. For the *n*th place (*topos*) in the periodic table there is exactly one elemental type, the one whose tokens have exactly *n* protons in their nuclei. These nuclei may or may not have the same number of neutrons. Often they do not. For instance, naturally occurring instances of the nineteenth element, potassium (K), may have 20, 21, or 22 neutrons; and those of the first, hydrogen, may have 0, 1, or 2 neutrons. Physically, these are allotopes; their nuclei and therefore the atoms themselves differ in mass and other physical properties. The mass of deuterium ($^2H$), for instance, is nearly twice that of protium ($^1H$). Chemically, however, their structures are nearly the same; even deuterium atoms react in much the same way protium atoms do.[33] Accordingly, in 1923 chemists decided to regard these different sorts of atoms as isotopic, and to identify *belonging to the same element* with *having the same number of nuclear protons*.[34] Their decision was a reasonable one, but as ontologists we wonder whether $^2$H and $^1$H are not really allotopes belonging to different natural kinds.

Third, there is the decay problem. Some isotopes are radioactive. For instance, $^3$H (tritium) decays into $^3$He (an isotope of helium). Chemically, hydrogen and helium differ greatly in kind, so when $^3$H atoms become $^3$He atoms they must undergo substantial change. This makes sense, for although the mass before and after are nearly the same, the atom before and the atom after are very different: the latter is more stable than the former, it is neither isomeric nor isomorphic with the former, and it is much less reactive. But why does the transformation of a neutron into a proton (with or without the addition of another electron) count as a substantial change for the atom, whereas the addition of a neutron to its nucleus or the loss of an electron from its outermost shell does not? Physically, such transformations seem no less substantial.[35]

The potentiality, isotope, and decay problems all raise the following question: With respect to the basic tokens of elemental kinds, does chemistry or physics offer the more accurate account of them as they are in their natures? We have at least two reasons for doubting that it is chemistry. First, there is its pragmatic character, which is manifest in the efforts of chemists to synthesize new materials and put them to use, in their willingness to regard material things as substances, their belief that rough and ready descriptions of molecular and atomic structure are good enough, their eagerness to exploit quantum physics without worrying too much about the theoretical implications, and in their disinclination to take seriously the metaphysical questions we are asking. Perhaps, then, as David Wiggins and Jaap van Brakel have suggested, their preferred stereotypes describe pragmatic rather than natural kinds.[36] Second, chemists seem in retrospect to have established their subject matter by focusing on supervenient properties of atomic and molecular systems and *deciding* to regard these properties as essential. They agree that the natural world is the physical world, that the natures of things are to be discovered by attending to their micro-constituents, and that the ultimate constituents of matter lie just beyond the scope of their discipline. Following this lead brings us to particle physics.

## FROM PARTICLES TO FIELDS

Of elementary particles and their interactions the foremost theory today remains the Standard Model. It divides its subject matter into matter and exchange particles. All matter particles have an intrinsic spin of $\frac{1}{2}\hbar$ and are thus fermions. Elementary fermions are of two sorts: quarks, fractionally-charged particles that participate in the strong force and constitute hadrons, and leptons, which do not participate in the strong force and do not constitute hadrons. Quarks and leptons alike are divided into three families, called "generations," each divided into two specific kinds. So there exist just twelve kinds of elementary particles constituting the material world. We can set them out neatly in a single table.[37]

**Table 6.1.**

| Generation | Flavor | Charge | Mass | Lifetime |
|---|---|---|---|---|
| | u (up) | +2/3 | ~3 | |
| | d (down) | -1/3 | ~5 | |
| 1 | $e^-$ (electron) | -1 | 0.510999 | ∞ |
| | $v_e$ (e-neutrino) | 0 | [$<0.5 \times 10^{-6}$] | ∞ |
| | c (charm) | +2/3 | ~1270 | |
| | s (strange) | -1/3 | ~101 | |
| 2 | $\mu^-$ (muon) | -1 | 105.658 | $2.19703 \times 10^{-6}$ |
| | $v_\mu$ ($\mu$-neutrino) | 0 | [$<0.5 \times 10^{-6}$] | ∞ |
| | t (top) | +2/3 | 172,000 | |
| | b (bottom) | -1/3 | 4,300 | |
| 3 | τ (tau) | -1 | 1,776.82 | $2.91 \times 10^{-13}$ |
| | τ (τ-neutrino) | 0 | [$<0.5 \times 10^{-6}$] | ∞ |

Once again, those who would be natural-kinds realists, and who look to natural science for metaphysical guidance, could hardly ask for more.

Yet things are more complicated than table 6.1 suggests. For while there may be only three generations of quarks and leptons,[38] each of the six quark-types comes in three colors—red, green, and blue—which physicists regard as distinct species[39]; and to each of the of the twelve quark- or lepton-types there corresponds an antiparticle type—to the up quark the anti-up quark, to the electron the positron, etc.—which exist no less than the anti-types with which they are paired.[40] So there are not twelve specific kinds of elementary particle, as our table suggests, but forty-eight.

Members of these species are all the *matter* particles or fermions. But there are also elementary *exchange* particles, or bosons, each having an intrinsic spin of $1\hbar$, which fall into diverse kinds depending on the forces they manifest: photons, the electromagnetic force; gluons, the strong force; W and Z particles, the weak force. There is one species of photon, eight species of gluon, and three species of W/Z particle. That makes twelve bosonic species in all, or thirteen if we count the Higgs boson, which the Standard Model requires in order to explain the masses of $W^\pm$ and $Z^0$ particles.[41] Should we then count sixty-one ultimate species or only forty-eight? To this question physicists offer two answers: (1) "The ultimate constituents of matter are fermions, to which the Pauli Exclusion Principle applies. Although fermions interact only through the mediation of bosons, the interactions are of fermions with fermions. Bosons, to which the Pauli Exclusion Principle does not apply, constitute the force-fields produced by fermions, and are not on an equal footing with them." (2) "The ultimate constituents of nature are fermions together with bosons. All interactions are to be understood in terms of fermion-boson couplings. Although very different in kind, elementary fermions and elementary bosons are on an equal footing; neither is any more particle-like or wave-like than the other, and each may give rise to the other." Answer (1) favors the forty-eight count, but is more suited to classical electrodynamics and our everyday conception of the world.[42] Answer (2) is better suited to quantum field theory, to the Standard Model in particular, and it favors the sixty-one count.

In any case, particle physicists take flavor-types to be real. In practice, moreover, they take (real) tokens of these types to be enduring things, each of them changing in some respects while remaining unchanged in others, as long as it exists. For practicing physicists, then, (real) quarks and leptons are individual continuants, and the specific or generic types they exemplify are well-defined universals. These are the fundamental stereotokens and stereotypes in natural science today.

To define an elementary particle, one must distinguish its state-independent properties from those which are state-dependent. The state-independent properties of an electron include its charge, (rest) mass, spin, and lepton number; its state-dependent properties include its particular location, momentum, total energy, and whatever spin-orientation it may exhibit in a Stern-Gerlach experiment.[43] An electron varies with respect to the latter while remaining invariant with respect to the former, as long as it exists. And a particle instantiates *the electron,* we may say, just in case its state-independent properties instantiate these same properties taken universally. A complete list of these universals is the stereotype associated with the term "electron." If particles each have just these intrinsic properties, they are necessarily electrons.

A happy result for scientific essentialists! But there is something paradoxical about it, for the members of classes so defined are usually abstract objects stipulated within a theory.[44] Take, for instance, the rectangular cones of Euclidean geometry. They are exactly alike in all essential respects, and we know just what these respects are by inspecting the appropriate definition.[45] Elementary particles are a lot like these geometrical solids: members of the specific kind K are essentially "identical"; and something is either a member of K or not a member of K, there are no borderline cases. Some philosophers have concluded that elementary particle-types are abstract rather than natural, and that the apparent necessity of their intrinsic properties is semantic rather than metaphysical.[46] With respect to the first point, at least, particle physicists disagree; for them, the particle-types posited in their theory are abstract *and* natural. The Standard Model itself is a marriage of the physical with the mathematical. Newton would have applauded it, however surprised he might have been that nature turns out to be fundamentally heterogeneous. And as for the second point, it would be safer to say that in the case of such objects the distinction between semantic and metaphysical necessity is very difficult to make.

But this theory is not without taxonomic complications. First, there is the virtuality difficulty. In any Coulombic interaction between two electrons, for example, it may happen that a photon being exchanged materializes into an electron-positron pair that quickly dematerializes into a photon. Such particles are very short-lived and indeterminate in their momentum and energy.[47] They are "virtual," physicists say, whereas the two electrons we started with are "real." And whereas the mass of a real electron is determinate and unchanging, "on mass shell," that of a virtual electron fluctuates indeterminately; it is "off mass shell." But then it is not an electron, since it fails to meet one of the conditions mentioned in the stereotype. Perhaps we should say that virtual electrons are quasi-electrons, or that they would be electrons if they were on mass shell; but then the class of electrons is not as well defined as Table 6.1 implies. Or perhaps we should say that virtual particles are useful fictions. Certainly they are unobservable. Yet physicists take them to be real in the sense that without them some phenomena could not be explained.[48] For real electrons, moreover, being involved in the electromagnetic field is a non-eidetic condition of manifestation. Consequently, (1) any electron interacts with itself via virtual photons, so that frequently it is a little off mass shell; the on-shell mass assigned to it by the stereotype is a mean value, the "effective" mass.[49] And (2), about any electron there is always a cloud of virtual electron-positron pairs that screen and diminish the strength of its electric charge.[50] It is tempting to think of the real electron as the one stripped of its virtual penumbra, but there is no such entity. The bare electron is an ideal type. The real electron is the whole package, the point-like core together with its atmospheric spread—the "effective" electron.[51] But *these* particles are neither incomposite nor structureless, and the usual accounts of them are grossly inadequate. It is misleading even to describe an existing electron as its "virtual" cloud *plus* its "real" core.

The mixed-state difficulty presents a second complication. Two examples will suffice. (1) There are experimental situations in which neutrinos appear to be of three distinct kinds.[52] When produced by the Sun and moving "freely" through space, however, they appear to be in a single superposition state which oscillates as it evolves, so that most of them are at one time "electron-neutrinos," at a later time "muon-neutrinos," and so forth.[53] It follows that they are not massless after all, and that lepton number is not always conserved. Does it not also follow that neutrinos belong to just one natural kind, a kind that has three distinct "isotopic" manifestations?[54] (2) Electromagnetic and weak interactions are united in the sense that both are describable by equations exhibiting the symmetry group SU(3)⊗U(1), which have four possible solutions, each representing a distinct way in which the electro-weak force can manifest itself, namely, as an isotriplet of vector bosons W and an isosinglet B, all in a "mixed state."[55] Now in fact, as the universe cooled, this symmetry broke spontaneously; the electro-weak field divided into the weak field with its massive $W^+$ and $W^-$ and $Z^0$ particles, and the electromagnetic field with its massless $\gamma$ particles (photons).[56] These particles seem very different in kind and are regarded as such in practice, but in theory their heterogeneity seems to mask the symmetrical state in which they originally existed. So which is it? Did the breaking of this symmetry bring to light four natural kinds or did it hide the one natural kind of which these are now four low-energy varieties? Should taxonomists take their bearings by the phenomenological and historical or by the ahistorical and theoretical?

Third, there is the decay difficulty. The Standard Model is a quantum field theory, and quantum field theories, unlike classical quantum mechanics, allow for the "creation" and "annihilation" of particles. It thus seems to admit our distinction between substantial and nonsubstantial change, and physicists are confident they can make this distinction when dealing with specific cases. We, however, should be more circumspect. So, for instance, where quantum field theorists speak of creation and annihilation operators, let us speak instead of transformation operators, thus leaving open the question of how radical these changes are. And to decide which transformations are truly substantial, let us first give ourselves a range of permissible cases. I offer the following: (1) An electron emits a photon and is deflected from its former path; (2) a blue up quark emits a gluon, becoming a red up quark; (3) a down quark emits a $W^-$ boson, becoming an up quark; (4) a muon emits a $W^+$ boson, becoming a muon-neutrino; (5) a strange quark emits a $W^-$ boson, becoming an up quark; (6) an electron-positron pair becomes a $Z^0$ boson; and (7) a $Z^0$ boson becomes a $\pi^0$ meson. Which if any of these are cases of genuine annihilation or creation (*genesis*)?

Evidently (1) is not a case of substantial change, even though the electron has come to have a different momentum and its total energy is diminished. Cases (6) and (7) evidently involve substantial change; in the Standard Model, at any rate, the transformation of two leptons into a weak-field boson, or of a weak-field boson into a quark-field fermion, is too great to be merely attributive. Case (2) is more difficult, since what physicists call "species" of up quark may in truth be isotopic varieties of one specific kind, which they have chosen to regard as allotopic because color-charge is so important in their account of strong-force interactions. Perhaps, when trying to identify the *natural* kinds here, we should look to differences in *flavor*. This brings us to cases (3) and (4). Physicists usually regard them as cases of annihilation and creation, yet their theory allows us to regard them as cases of metamorphosis in which the particle's essential mu-ness or up/down-ness (understood as mixed states) is preserved.[57] But if flavor changes are nonsubstantial, then why not generational changes as well? Even (6) is not plainly a case of annihilation. Perhaps the fermionic species are leptonhood and quarkhood, just two in number, each of which comes in the six isotopic varieties displayed in Table 6.1. Perhaps, as Steven Weinberg and others have suggested, these differences in charge, mass, flavor, and generation all mask what is fundamental.[58] My argument here is of the slippery-slope variety, and one may slip and slide in either direction. The question is this: To what do we appeal when drawing the line? There is little in the things themselves to which we can appeal, since they are not the sort of thing to which our distinction between substantial and nonsubstantial change clearly applies. Elementary particles are not exemplary continuants. Are they continuants at all?

That elementary fermions are quite unlike ordinary material things becomes clear when one attends to cases such as these: (1) An electron's location and momentum are more or less indeterminate according to Heisenberg's uncertainty principle, hence it cannot have a determinate path and is therefore not the sort of thing that could be tracked. (2) When a neutron decays into a proton, not even God could tell which of the two down quarks in the former has been transformed into an up quark, nor could anyone tell which of the two up quarks in the latter has just been "created." (3) We shoot two electrons *a* and *b* directly at each other at high speed. A moment later we observe two electrons, *c* and *d*, coming back at us. Does $c = a$ and $d = b$? Perhaps, but we cannot rule out the possibility that $c = b$ and $d = a$, or that neither *c* nor *d* is the same as *a* or *b*. Reidentification over time seems impossible. (4) A helium atom has two electrons. They differ in spin orientation according to Pauli's exclusion principle, yet there is no way of telling which is which, of labeling or counting them. It is a case of cardinality without ordinality; and were the atom's energy increased to the point that it disintegrated, there would no longer be any way of telling how many electrons were present; their cardinality, too, would be gone.[59]

It is the pronounced wave-like character of elementary fermions that makes them so unlike ordinary things. They are as much wavicles as particles. So too, of course, are elementary bosons. Call them "quantum objects." Such objects fall naturally into discrete classes, each having many members, and yet its members are untrackable, indiscernible, and uncountable. But as P. Strawson, S. French, and others have argued, something is an individual only if it is countable, discernible, and trackable.[60] Thus quantum objects are not individuals; nor are they continuants of some other sort. For something is a continuant only if its situational involvement in the world is grade II. Quantum objects are normal modes of quantum fields.[61] And these modes, or field-quanta, depend essentially on the fields to which they belong. Hence, like Spinoza's finite modes (of the one substance that is nature), their situational involvement is grade III. Though real, inasmuch as they are the normal modes of real fields, they are not entities in their own right. Hence, contrary to what we initially thought, these fundamental stereotokens are not continuants and the stereotypes they instantiate are not natural kinds in the primary sense of the term. Particle physicists may go on regarding field-quanta as if they were enduring individuals, in order to facilitate the role of Feynman diagrams in calculating cross-sections, decay rates, or the like; but then these individuals, as such, are what I have called *pragmata* and the discrete classes into which they fall are pragmatic kinds.

We see, moreover, that for the Standard Model the primary *entities* are these quantum fields to which elementary fermions and bosons belong as normal modes. Let one such field be called "*f*." In its ground state, *f* is a Bacchanalian revel of virtual quanta in which no "real particle" exists.[62] To create one such object, increase the energy of *f* to level one; for two such objects, increase it to level two; and so forth.[63] Field *f* is an ever-active complex whole that is capable of interacting locally with other such wholes; no quantum field is indifferent to all other quantum fields. Those who would be natural-kinds realists, and who look to the Standard Model for guidance, should not let the pragmatic particle-kinds they have found blind them to the underlying fields and *their* kinds, for these alone appear to be the genuine stereotokens and stereotypes posited in today's quantum field theories. With respect to *these* stereotokens, two questions arise: What sort of entity are they, and how many of them are there?

It is hard to say what *f* is. If you focus on it as a whole you have trouble remaining faithful to its essential multiplicity, while if you focus on the fact that its interactions are all local, you have trouble remaining faithful to its essential unity.[64] In any event, quantum field theorists describe *f* as a four-dimensional occurrent, a perdurant rather than an endurant, and perdurants are not continuants, hence their types are not natural kinds in the primary sense.[65] It is also hard to say to say how many such fields there are. The Standard Model gives no clear answer. One may count as two the electron and photon fields, as physicists often do, and by proceeding in this piecemeal way end up with several dozen fields, each one of a kind. Alternatively, one may regard the electron and photon fields as two aspects of a single entity, and by proceeding in this way end up with comparatively few fundamental kinds. So how many are there?

The Standard Model commits us to the existence of at least two fermionic fields. But this is just one of several respects in which it is now thought to be inadequate. Physicists today are in search of a theory that is more unified, that takes into account the gravitational force and the early history of the universe, and that is more deductive. Table 6.1 resembles taxonomic schemes in nineteenth century chemistry. It presents a system of apparent natural kinds without saying *why* the things described are as they are—why the mass of an electron is what it is, why the coupling strength of the electric field is what it is, and so on.[66] It is thought to be insufficiently explanatory. It also describes the being of particle-types apart from their *genesis*—how there came to be leptons as distinguished from quarks, fermions as distinguished from bosons, and so on. Table 6.1, like Mendeleev's periodic table, is typological without being sufficiently historical; it takes as given what appears to be a contingent result. It is also parochial, having nothing at all to say about the dark matter which makes up most of the material universe.[67] Physicists seem justified, therefore, in seeking a more comprehensive theory. And if their anticipations of this theory are correct, it will likely commit one to a field ontology that is monistic rather than pluralistic.[68] That it will not be hospitable to natural-kinds realism becomes clear when we glance at three recent developments.

*From Heterogeneity to Homogeneity: Superstring Theory*

In the electroweak theory, particles or fields which seem fundamentally different in kind appear to be low-temperature "solutions" masking a still more fundamental symmetry. The same is true in grand unified theories (GUTs), which unite the electromagnetic and weak fields with the strong field, thereby superseding the differences between quark and lepton.[69] Superstring theory takes this reductive tendency one step further: by superseding the difference between quantum mechanics and general relativity theory it can account for the gravitational force as well. The primary "things" are no longer point-like particles but one-dimensional "strings." These objects are like very tiny loops, on the order of 10–$^{33}$ cm in length, vibrating incessantly in nine-dimensional space. The vibrations of these strings may occur in many different patterns or modes. In one mode they appear to be electrons with their characteristic spin, mass, and charge; in another mode, up quarks with *their* characteristic spin, mass, and color-charges; in yet another, gluons; and so forth.[70] All such differences in kind are only differences in state, and the entities in these different states are all the same in kind. Hence there is just one natural kind—or not even one if, as it seems, these strings are really many-dimensional perdurants. Again we must distinguish what the theory says from what the theorists say when speaking informally.

Superstring theory is a highly speculative work in progress. It may need ten spatial dimensions rather than nine, and some additional objects.[71] Eventually it may be rejected altogether. Yet it represents the efforts of quantum field theorists to understand everything universally in terms of a few equations exhibiting timeless symmetry principles and having various possible solutions; and insofar as this outlook prevails in physics, the concept of a natural kind will appear to have little or no theoretical role.

*From Being to Becoming: An Evolutionary Model of the Universe*

Fundamental physics today comprises quantum field theory and cosmology. The former tends to regard time as symmetrical, space-like; it features equations some of whose solutions describe the actual universe. The latter tends to regard time as asymmetrical; it features an historical account of the actual universe, from its origin in a "Big Bang" to the current "Big Freeze." Between these two fundamental disciplines there is some tension. The problem, apparently, is to understand together being with becoming, the time-symmetric invariance described by quantum field theory with the time-asymmetrical variation described by cosmology. We have already encountered this problem in a very different context.[72] The prevailing tendency among physicists is to privilege timeless being. Yet some claim that irreversible becoming is basic. Both sides agree that our universe originated in a big bang, followed almost immediately by inflation, and that as it cooled there occurred a sequence of spontaneous symmetry-breaking (speciation) events, in the course of

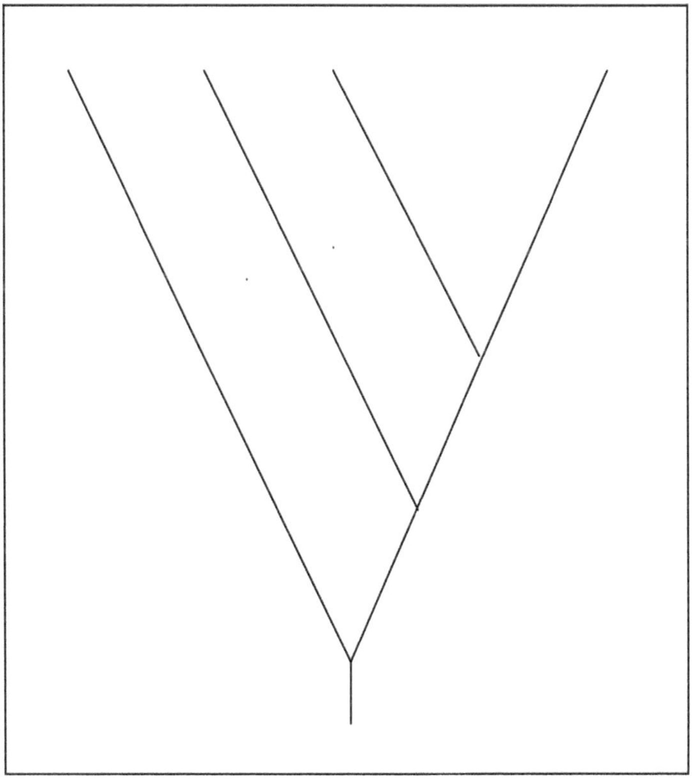

**Figure 6.1.**

which one force became four. This process, considered as irreversible or not, can be represented by a tree-like diagram (figure 6.1). The idea is an old one. In this, its latest iteration, "the many" is our world in its cooler, more developed state, while "the one" is this same world in its primordial unity, before any of these phase transitions had occurred.[73] So far, then, physicists agree. But a few cosmologists have gone much further, claiming that our world is but one of many "bubbles" in a multiverse of worlds, each with its own laws and history. The generations of worlds within worlds, in a multitude of big bangs, can be represented by the same tree-like diagram. *Our* world, then, is just one of the many contingent "twigs," each with its own quantum fields, its own coupling constants, and its own physical laws. The laws that field theorists today take to be both necessary and universal are, in fact, neither.[74]

The similarity between cosmogeny and phylogeny suggests an evolutionary model of the multiverse. How far should this analogy be pushed? Lee Smolin, for one, pushes it pretty far.[75] A universe containing black holes, he argues, can produce other universes. If it does, the descendants will differ slightly from one another and from their common ancestor. These mutations are heritable and cumulative over generations. Some populations (species) of universes are also more fit than others in the sense that their members, having more black holes, are more productive of universes like themselves. Hence some "species" will remain rather small while others proliferate dramatically. But how illuminating is the analogy with natural selection? Is the multiverse evolving irreversibly? And if so, is its evolution not governed by any overarching laws? Are there no all-pervasive quantum fields or quantum types?[76] In any event, if we suppose that our world does contain genuine continuants belonging to discrete natural kinds, we have no reason to think they exist in any world other than our own.

### *From Things to Relations: Structural Realism*

We ordinarily take singular things to be prior to relations. In our focus on enduring individuals in particular, relations among them tend to disappear. Call this everyday way of conceiving the world "individual realism." Theoretical physicists, to the contrary, take relations to be prior to objects so related. In their focus on invariant laws or structures, both of which are essentially relational, *things* tend to disappear. Philosophers call this scientific way of conceiving the world "structural realism."[77]

With respect to the ultimate *relata* of real relations, structural realists have the following options. (1) Unqualified realism: these *relata* exist and are entities in their own right. (2) Unqualified antirealism: there are no such things. (3) Qualified realism or antirealism: they exist, but only as field quanta or the like, not as entities in their own right. (4) Agnosticism: their ontic status lies beyond the scope of scientific theory. Most structural realists accept (2), (3), or (4) and reject (1). We might think that a holistic version of (1) is to be preferred, since there could be no structure without that which is structured and if, as it seems, the mathematical structure of the world is itself is a universal that the physical world instantiates, it would seem to follow that there is some underlying substance in which this spatiotemporal structure inheres. But there is no such entailment; we need not admit any distinction between the world itself and that to which its structure belongs as a property.[78]

Shifting between individual and structural realism involves a conceptual Gestalt reversal. The two outlooks seem mutually exclusive. Perhaps they are complementary in Bohr's sense of the term.[79] Yet Brian Ellis and other scientific essentialists today believe that the two can be combined, that we can have both individual continuants exemplifying universal types *and* physical laws exhibiting universal symmetries. To examine this belief is our next task.

## NATURAL KINDS AND LAWS OF NATURE

The Aristotelian tradition of natural kinds focused primarily on human beings, horses, and the like. Such things were thought to be entities (substances), some of them belonging to discrete species in virtue of their inner natures. In this tradition the key metaphor is *eidos,* a term ordinarily associated with seeing (knowing) and what can be seen (the forms of things). Exemplary kinds include *humankind, horses,* and *water,* but not *bodies* or *movables.* By contrast, the Galilean tradition of natural law focused primarily on the motions of falling bodies, pendulums, and the like. Such occurrences were thought to exhibit universal patterns in virtue of their inner connectedness. Here the key metaphor is *lex,* a term ordinarily associated with rules governing or regulating behavior in a civil society. Exemplary laws include $S = \frac{1}{2}gt^2$, $T = 2\pi(l/g)^{\frac{1}{2}}$, $m_1v_1 + m_2v_2 = m_1v_3 + m_2v_4$, and $F = ma$. They do not include exemplary definitions such as *Man is a bipedal animal* or *Bodies are delimited quantities of matter.*

As in chapter 5 we regarded exemplary "species" as sortal concepts which may or may not denote natural kinds, so now let us regard exemplary "laws" as propositions which may or may not describe laws of nature. These propositions are equations in which the values of some physical quantities are held to be functions of the values of other physical quantities. Functions are relations whose relata—variables to which real numbers can be assigned—are so connected that one is dependent on another as a matter of mathematical necessity: if $y = x^2$ and $x = 2$, $y$ must equal 4.[80] Laws of nature are those invariant relations whose relata—varying physical states to which real numbers can be assigned—are so connected that one is dependent on another as a matter of natural necessity: if $S = \frac{1}{2}gt^2$ and $t = 2$ seconds, $S$ *must* equal 2g in length.[81/82] In chapter

5 we asked what it is to be a natural kind and whether there are any such kinds. Now let us ask what it is to be a law of nature and whether there are any such laws. My answer must be very brief. I set it out in a pair of claims, each in need of considerable elaboration and yet plausible enough, I trust.

First Claim: Physics has moved well beyond its origins in classical mechanics and electrodynamics. It now features 4-vectors, complex numbers, matrices, and operators, as well as more traditional functions. It is also about action more than locomotion, phase and potential more than velocity and force.[83] Yet even without going into these details we can say that a law of nature is an invariant universal relation (connection, nexus) whose relata are quantifiable event-types and whose proper representation is an equation or the like. The universality condition excludes relations in nature that are merely local but allows for laws concerning interactions that are everywhere local. The invariance condition excludes relations that are merely accidental but allows for laws that are not causal or deterministic, and that are only contingently necessary if the universe itself is contingent.[84] The representation condition excludes relations of mere resemblance and those which are merely ordinal,[85] but allows that Heisenberg's uncertainty relations may be laws.

Many so-called laws of nature are not strictly laws. One could argue, for example, that a law meets the invariance condition only if it is exceptionless, only if the relation itself remains the same no matter how the relata *and* their particular circumstances may vary.[86] Hence one should not expect to find many genuinely nomic statements beyond fundamental physics. Perhaps there is just one law of nature, to be represented by an equation for the entire universe—though no one has determined what that equation might be. Does it follow that the concept just now delineated is too restrictive? Not at all. We may allow for the application of this concept to relations that are not strictly laws of nature, provided they are sufficiently law-like. I would argue that reaction equations in chemistry may be regarded as nomic if they are correct, and so too some statements of the *ceteris paribus* variety. But the statement, "The electron's mass is 0.510999 GeV/$c^2$," should not be regarded as nomic even if having that mass is a necessary property of electrons; the *nomic* statement in this case would be an equation or set of equations from which the electron's specific mass falls out as a solution.[87] In making such decisions, the question is always whether the analogy proposed is appropriate: Are the relations in question sufficiently like laws in the primary sense of the term that it would be illuminating to classify them as laws in some secondary sense of the term? Linguistic propriety is another matter. Scientists use the words "law," "equation," "principle," "model," and "theory" in ways that often seem rather muddled, but there need not be anything seriously wrong in that.

Second Claim: On the question of whether laws of nature exist apart from our legislative proclivities, four options are available. (1) "There exist relational universals of the right sort, at least one of which is exemplified in necessary connections among events in the actual universe." This is unqualified realism with respect to laws of nature. It comes in two main varieties, depending on whether one takes the relations in question to be external or internal to their relata.[88] (2) "There are necessary connections among events in the actual universe but they do not instantiate nomic universals; there are no such universals." This is qualified realism, if indeed our law-like idealizations can be well founded in reality. Stephen Mumford, however, has highlighted its antirealist character: nature is lawless, he argues, even though physical properties are essentially dispositional and internally interrelated.[89] (3) "There exist nomic universals but none of them is exemplified in connections among actual events." One might regard this position too as qualified realism, but I would emphasize its antirealist character: nature is lawless unless laws of nature are somehow *in* it. (4) "There exist no nomic universals and no necessary connections among actual events." This is unqualified antirealism with respect to laws of nature. Its proponents may nonetheless argue that law-like statements or equations can be useful in explaining and predicting phenomena.[90] Which of these four options is the correct one has not been settled. And supposing it is the first, how do we discover what the laws are? The resources available to us are conceptual as well as empirical. Philosophers interested in this issue should pay more attention to the way in which physicists use symmetry principles, or Lorentzian (global spatiotemporal) and gauge (local phase) transformations, to determine the fundamental laws.[91]

Now compare natural-law realism, which most scientists accept, with natural-kinds realism. The latter requires s-resemblance classes of continuants. The invariance to which it looks is of enduring things amid qualitative, quantitative, and relational changes. It does not require necessary connections among events, and may take nomic relations (scientific as well as nonscientific) to be but shadows cast upon the world by our discursive minds. Given its focus on heterogeneity, it probably denies that nature is fundamentally homogeneous or that there is just one natural kind. Natural-law realism, by contrast, requires necessary connections among events. The invariance to which it looks lies here, in these very connections. It does not require continuants and may take all stereotypes (scientific as well as nonscientific) to be but shadows cast upon the world by our pragmatic interests. Given the comprehensiveness of its outlook, it probably does not rule out the possibility that nature is fundamentally homogeneous or that all real events are governed by just one natural law.

Natural-law realism and natural-kinds realism present very different outlooks on the world, but they are not formally contradictory. Following Hume, we may reject both, on the grounds that inner natures and inner connections are posited by us in order to tidy up the messiness of the phenomena and feed our desire to feel somewhat in control. Or, following today's scientific essentialists, we may accept both. Suppose we do. Then we must say whether we accept qualified or unqualified realism with respect to natural kinds and laws of nature.[92] We must also say how (if at all) natural kinds and laws of nature are connected, and to what extent (if any) natural kinds in the primary sense correlate with laws of nature in the primary sense. Here lie grave difficulties, as we can see by glancing at the works of two scientific essentialists.

E. J. Lowe posits four categories of being: substantive universals (kinds), nonsubstantive universals (properties), substantive particulars (objects), and nonsubstantive particulars (modes). Properties are instantiated by modes, which characterize objects. Kinds are instantiated by objects and characterized by properties; and it is among these kind-characterizations that the laws of nature are to be found. As exemplary nomic statements Lowe mentions "Benzene is flammable," "Common salt is water-soluble," "Gold is soluble is aqua regia," and "Uranium explodes when it exceeds its critical mass." But even if these generalizations are necessarily true, they are not *nomic*. Lowe suggests that his account can be extended to Newton's law of gravitation construed as "a fact about the nature of massive bodies quite generally." "The law," he says, "is that masses of matter attract each other with a degree of force which is proportional the product of the quantities of their mass and inversely proportional to the square of the distance between their centers of mass." But *mass* and *gravitational force* are not exemplary natural kinds, and Newtonian mechanics does not require that tokens of either kind be irreducibly thick particulars. Nomic statements cannot be assimilated to statements of essential predication whose subject terms designate substantive universals. To combine natural-law realism with natural-kinds realism one must radically modify if not abandon the Aristotelian categorial scheme that is Lowe's starting point.[93]

Brian Ellis also posits powerful particulars and the substantive universals (natural-kind types) they instantiate, as well as dispositional properties and the property-universals *they* instantiate. Some dispositional properties belong essentially to the things that have them, and it is in these properties that Ellis grounds both natural kinds and laws of nature.[94] One such property is electric charge Q. On the one hand, some specific tokens of Q belong necessarily to electrons, those powerful particulars which are tokens of the universal *electron*. On the other hand, tokens of Q interact with one another necessarily in accordance with Coulomb's Law: $F = q_1q_2/r^2$. That electrons have specific charge = -1, and that the electric force repelling one electron from another varies inversely as $r^2$, are a pair of *a posteriori* necessities. Consequently, from the natural kind *electron* or the dispositional property Q one cannot deduce Coulomb's Law, just as from Coulomb's law one cannot deduce the existence of enduring particles having electric charge = -1.[95] We could imagine a world in which $F = q_1q_2/r^3$, but it would not be one in which there could be any electrons or any instances of the property-universal Q.[96]

Is this true? Maybe. Yet there are difficulties, three of which bear on our present theme. (1) Ellis' account of the world is rather classical. Exemplary for him are such laws as those expressed by reaction equations in chemistry.[97] They mention natural kinds in the primary sense, it seems, but what they express are not laws of nature in the primary sense. Coulomb's Law does express a law of nature in the primarily sense, it seems, but requires only that there be tokens of Q, not that these tokens be properties of irreducibly thick particulars. Ellis takes electrons to be enduring particles which reach out in all directions, as it were, in the electric force fields they generate, and which differ in kind from other charged particles (protons, quarks, muons, $W^-$ bosons, etc). Apart from the quantization of electric charge, his account of the electric field remains Faradayan. It is not the account we find in quantum field theory.[98] (2) When he does focus on the most fundamental laws, he notices that many of them are structural rather than dispositional, global rather than local, inasmuch as they pertain to the universe as a whole, and that they do not to require natural kinds in the primary sense.[99] To square this fact with his natural-kinds realism, Ellis takes the universe itself to be one of a kind.[100] But this kind need not be substantive, since any instance of it need not be substantive or of something substantive. And if the universe is one substance, there can be no multitude of substances within it. He should then give up his pluralism with respect to (instantiated) natural kinds.

Difficulties (1) and (2) arise from the fact that natural science, on which Ellis depends, harbors an unresolved tension. On the one hand we have quantum field theories. They are fundamental, more theoretical than phenomenological, and feature equations which may well express laws in the primary sense. They do not describe a dappled world in which there are manifest asymmetries, real causes, or continuants and their kinds. On the other hand we have the special sciences. They are less fundamental, more phenomenological, and feature stereotokens and stereotypes which may well count as natural kinds in the primary sense. They do not describe a symmetrical world in which there is real homogeneity, quantum fields, or laws of nature in the primary sense.[101] Again we face a dilemma: natural kinds without laws of nature or laws of nature without natural kinds. Ellis' scientific essentialism mirrors this dilemma without resolving it.

There is yet another, related difficulty: (3) From its beginnings in the seventeenth century, natural science has tended to reduce the more phenomenological to the more theoretical, the derivative to the fundamental—which means in effect to reduce the biological to the chemical, the chemical to the physical, the heterogeneous to the homogeneous, the classical world of enduring things and their kinds to the quantum world of fields whose interactions are governed by laws. To preserve their natural-kinds realism in the face of this reductionist tendency, scientific essentialists must show that a theory of everything could not be a theory of every *thing*. And to show this, they must give reasons for believing that enduring things (continuants) have in fact emerged out of interacting quantum fields. Enduring things are irreducibly thick particulars (substances), whereas field quanta are not. Scientific essentialists must therefore give reasons for thinking that substantive emergence has occurred. Ellis, for one, has not done so. In this respect, therefore, his natural-kinds realism remains in need of a defense. I shall outline such a defense in the final section of chapter 7. Before doing so, however, let us examine some of the stereotokens and stereotypes to be found in the biological sciences.

## NOTES

1. Mill 1973, I.vii.4.
2. Peirce 1940, I: 600f.
3. Kripke 1980, 71ff; and H. Putnam, e.g., "The Meaning of 'Meaning,'" in Putnam 1979, 215ff.
4. Putnam, "Is Semantics Possible?," "Explanation and Reference," and "The Meaning of 'Meaning,'" all in Putnam 1979, 147–52, 196f, 204f, 230, 236f, 247–52, 269, and "A Problem about Reference," in Putnam 1981, 22–48. Compare Searle on the use of "standard" or "identifying" descriptions in cases of singular reference, J. Searle, "Proper Names," *Mind* 67 (1958): 166–73. On the relation between Putnamesque stereotypes and Fregean senses, see: Wiggins 1980, 79, and Salmon 1981, 12, 153.
5. On the way in which samples fix reference, see: Kripke 1980, 138f, and Putnam, "Is Semantics Possible?," "Explanation and Reference," "The Meaning of 'Meaning,'" and "Language and Reality," all in Putnam 1979, 140–142, 197f, 223–28, 234–36, 250, 269f, 277, 283. They admit that the putative extensions, too, may change, see: Kripke 1980, 138f, and Putnam 1979, 142, 225. But then why think that the reference of a natural-kind term is any more fixed that its associated stereotypes? Here Kripke and Putnam have recourse to a causal theory of meaning transmission according to which the reference of "water," for instance, is fixed once and for all by some initial "baptism." This theory has been much criticized.
6. Kripke 1980, 97, 127ff, 159 with 53, 108–10.
7. Putnam, "The Meaning of 'Meaning,'" and "Language and Reality," both in his 1979, 227–29, 233, 274ff.
8. For how this might be done, see chapter 4 of this volume. That scientific essentialism cannot be derived within a theory of reference, see: N. U. Salmon 1981, 161–218. S. Mumford rehearses Salmon's critique as part of a more comprehensive argument against scientific essentialism: Mumford 2004, 113–16, 163, and context.
9. Cf. e.g., K. Fine 1975, 17–32; K. Donnellan 1983, 84–104; T. Kuhn 1979a, 459–82; L. Poncinie 1985, 415–27; I. Hacking 1988, 78–91; J. LaPorte 2004, 106f; H. Beebee and M. Sabbarton-Leary, "On the Abuse of the Necessary A Posteriori," in their 2010, 164–67.
10. See chapter 4 of this volume.
11. That "$H_2O$" is here used descriptively rather than referringly, see: Salmon 1981, 99; and Bealer 1987, 300f, 315, 361n21. Kripke's two errors are related. Taking them together, we see that by assimilating "Water is $H_2O$" to sentences such as "Hesperus is Phosphorus" and "The stuff in this drinking glass is the same as the stuff in that bathtub," he has in effect classified it with sentences such as "Hesperus is Venus," and

"The stuff in this glass is water." This makes it a *proto*metaphysical statement, which (if true) conveys us to something in the real world, rather than a *nature* statement, which (if true) tells us that thing is in its nature or essence. According to the theory they are defending, "water" and the like have already anchored us in the real world; scientists then discover what the realities so named are essentially, in their natures.

12. M. Johnston classifies "Water is $H_2O$" as a constitution statement in his 1997, 573, 578, 580–83. This is not incorrect, since in chemistry at any rate the nature of a thing is understood in terms of its (micro)constitution as well as its (micro)structure.

13. Cf. Mill 1973, 114f, 121f, 127.

14. Pauling 1970, 1.

15. Cf. Abbott 1997, 311–19.

16. Pure water at room temperature contains these ions in concentrations of 1x10–7 moles/liter. Cf. Pauling 1970, 482; F. H. Stillinger, "Theoretical Approaches to the Intermolecular Nature of Water," *Philosophical Transactions of the Royal Society of London* B278 (1977): 107; and P. Atkins 1994, 293.

17. Cf. Zemach 1977, 120–22; Mellor 1977, 303, 307–09; LaPorte 2004, 94–102; Weisberg 2006, 337–45. H. Putnam has acknowledged some of these difficulties, see: Putnam 1992, 216n8.

18. Cf. Needham 2000, 13–21; Brakel 2000, 71ff, and 2000a, 162–84.

19. For details, cf. D. Eisenberg and W. Kauzmann, *The Structure and Properties of Water* (Oxford: Oxford University Press, 1969); Pauling 1970, 439–43; F. H. Stillinger, "Water Revisited," *Science* 209 (1980): 451–57; F. Franks, *Water: The Matrix of Life* (New York: Springer, 2000), 9–53; J. Walker, "Boiling and the Leidenfrost Effect," in D. Halliday and R. Resnick, *Fundamentals of Physics,* 3rd ed. (New York: John Wiley & Sons, 1988), essay E10.1–6; M. C. R. Symons, "Water Structure: Unique But Not Anomalous," *Philosophical Transactions: Mathematical, Physical and Engineering Sciences* 359 (2001): 1631–46; O Mishima and H. E. Stanley, "The Relationship between Liquid, Supercooled, and Glassy Water," *Nature* 396 (1998): 329–35.

20. Pauling 1970, 339, and Timmermans 1963, 36–47.

21. Pauling 1970, 155. For $H_2S$ molecules, the bond-length and bond-angle are 133.6 picometers and 92.1 degrees, whereas for $H_2O$ molecules they are 95.8 picometers and 104.5 degrees.

22. Cf. R. G. Woolley, "Must a Molecule Have Shape?," *Journal of the American Chemical Society* 100 (1978): 1073–78, and "Molecular Shapes and Molecular Structures," *Chemical Physics Letters* 125 (1986): 200–05; Z. Jenkins 2003, S1053–S1057; but compare Ramsey 2000, 117–28. Ball-and-stick representations display invariant structure by reifying the mean values of several variables pertaining to the molecule when it is in a comparatively low-energy high-equilibrium state.

23. In making these computations, chemists rely on either the valence bond (VB) theory or the molecular orbital (MO) theory. For an introduction to both, see: Atkins 1994, 461–77, and, for a summary comparison, Atkins 1991, 236f. See also H. Primas, *Chemistry, Quantum Mechanics and Reductionism: Perspectives in Theoretical Chemistry* (Berlin: Springer, 1983); Scerri 1994, 160–70; Ramsey 1997, 233–51; Woody 2000, S612–S627.

24. Vibrations follow necessarily from the wave-character of all quantum objects. On resonance structures in particular, cf. Pauling 1970, 170–72, 231–35, 239–41; Atkins 1991, 309–11; Vermeeren 1986, 273–90.

25. In an $H_2$ molecule, neither electron can be allocated to one or the other proton exclusively. The MO theory emphasizes this fact and is thus more holistic than the VB theory. On the metaphysical implications of delocalization, cf. Needham 2006, 65–67, and 2006a, S829–S840; Scerri 2007, 114–17. The delocalization of electrons helps to explain electrical conduction. On the complexity of water in this respect, cf. H. J. Bakker and H.-K. Nienhuys, "Delocalization of Protons in Liquid Water," *Science* n.s. 297 (2002): 587–90.

26. On the importance of symmetry in molecular chemistry, cf. Atkins 1994, 12, 512ff, and 1991, 348–59. On the relation between symmetry and object- independence, cf. Earley 2006, 207–20, esp. 212f.

27. The difference is not merely one of size: a diamond may be a single molecule. The physical difference between intra- and inter-molecular bonds, or covalent and ionic bonds, or molecules and van der Waal's complexes, seems to be a difference of degree, whereas the chemical difference seems to be one of kind.

28. See chapter 5 of this volume.

29. Cf. P. Ambruster and F. P. Hessberger, "Making New Elements," *Scientific American* 287 (1998): 72–7; K. Stone, "The Element of Stability," *Science* 278 (1997): 571f; and Pauling 1970, 851.

30. See chapter 4 of this volume. Compare Bhushan 2006, 328–33. Yet Bhushan should nevertheless agree that samples of an element may come into being in any number of different ways. For example, lead isotopes may be produced in the radioactive decay of uranium or of thorium; and helium atoms may be produced by radioactive decay or by nuclear fusion.

31. Scerri 2007, 114–18, 278f; and 2005, 127–45.

32. This is the now-recommended term: IUPAC, Red Book 2005: *Nomenclature of Inorganic Chemistry*, section IR-5.3.2.2 http://old.iupac.org/publications/books/rbook/Red_Book_2005.pdf. Chemists do not take every free proton, regardless of its circumstances and history, to be a hydron.

33. The reaction rates differ and may differ markedly, as in the case of water and heavy water. This is why the latter, ingested in high concentrations, would kill us; our bodies could not use it quickly enough.

34. The decision was made by IUPAC. Cf. Scerri 2007, 160, 176ff, 316n3; H. Kragh, "Conceptual Changes in Chemistry: The Notion of a Chemical Element, ca. 1900–1925," *Studies in History and Philosophy of Modern Physics* 31B (2000): 435–50. Chemists are most interested in the outermost electronic shells of atoms.

35. The transformation described here is a case of $\beta^-$ decay, in which a down quark becomes an up quark and an electron with an anti-neutrino, or else a down quark and a neutrino become an up quark and an electron. See chapter 5 of this volume. Compare Ellis 2001, 237–41.

36. Cf. Wiggins 1980, 211f. J. van Brakel has emphasized the pragmatic character of chemistry in, e.g., Brakel 1997, 253–82, 2000, 72f, 84, et passim, and "The Nature of Chemical Substances," in Bhushan and Rosenfeld 2000, 162–84. Van Brakel does not seem to distinguish sharply, as I do, between pragmatic and natural kinds.

37. The electric charges are given in units of a proton's charge. The lifetimes are in seconds, the masses in GeV/$c^2$. Although the Standard Model takes neutrinos to be massless, the phenomenon of neutrino oscillation, known since the 1950s, requires that at least two of them have masses. They must however be very light. I have added in some rather conservative estimates. For more accurate data, see the particle listings page of the Particle Data Group at http://pdg.lbl.gov/2009/listings/contents_listings.html.

38. Cf. Griffiths 2008, 51; J. Allday, *Quarks, Leptons and the Big Bang* (Philadelphia, PA: Institute of Physics, 1999), 166 with 207; A. H. Guth, *The Inflationary Universe* (Reading, MA: Perseus Books, 1997), 128n.

39. Posited initially in order to preserve Pauli's exclusion principle, color has come to be regarded as a dynamical (dispositional) property analogous to electric charge. Thus color is to the gluon field what electric charge is to the photon (electromagnetic) field. Quarks have charge of both sorts.

40. A particle and its corresponding anti-particle may be oppositely charged—as are electrons and positrons—but this is not always the case. In neutrinos and antineutrinos, the difference must lie elsewhere.

41. The Higgs boson is a spin-0 particle, whereas the other twelve are spin-1. Physicists have now found that bosons of this type do exist.

42. Weinberg 1977, 22f; "What is an Elementary Particle?" in *Beamline 1996* http://www.slac.stanford.edu/pubs/beamline/27/1/27–1–weinberg.pdf.

43. On the Stern-Gerlach experiment, see: R. P. Feynman, R. B. Leighton, M Sands, *The Feynman Lectures on Physics* (Reading, MA: Addison-Wesley, 1965), III.5; R. I. G. Hughes, *The Structure and Interpretation of Quantum Mechanics* (Cambridge, MA: Harvard University Press, 1989), 2–8. Crucial here is the difference between a particle's ongoing eigenstate and the probabilistic eigenvalues associated with it in various experimental arrangements: Auyang 1995, chapter 4. I regard eigenstates as persisting dispositional properties, eigenvalues as their (necessarily probabilistic) occurrent manifestations. Of the electron's intrinsic properties mentioned here, all but one (its mass) are specified by "quantum numbers."

44. Note again the difference between Mill and Peirce on what a *natural* kind is. The members of Millian kinds are nothing like abstract objects, whereas the members of a Peircean kind may well be.

45. Cf. Zalta 1983, chapter 1. Whereas abstract objects "encode" their properties and are discovered axiomatically, he says, concrete objects "exemplify" their properties and are discovered empirically.

46. e.g., Mellor 1977, 307; and Khalidi 2009, 89. Compare, e.g., A Jaffe, "Where Does Quantum Field Theory Fit into the Big Picture?" in Cao 2004, 137.

47. These are consequences of Heisenberg's two uncertainty relations together with the fact that the spatial and temporal intervals involved in any such interaction are very small. So brief is the lifetime of the Δ+ hadron (on the order of ten to twenty-five seconds) that some physicists prefer not to classify it as a real particle. See J. Allday, *Quarks, Leptons and the Big Bang* (Philadelphia, PA: Institute of Physics, 1998), 136, 153n1. Should the same be said of W and Z bosons?

48. Consider, for instance, the Lamb shift or the Cassimer effect.

49. R. P. Feynman, Nobel Lecture, 1965 www.nobelprize.org/nobel_prizes/physics/laureates/1965/.

50. About any quark there is an analogous cloud, but their present confinement has an *anti*-screening effect; see D. Griffiths 2008, 69f, 300f.

51. Aitchison 1985, 354–58, 361f; and Auyang 1995, 189–93.

52. Noteworthy here is the work of L. M. Lederman and others in the early 1960s. Cf. D. Griffiths 2008, 29.

53. Griffiths 2008, 387–98; Aitchison and Hey 2004, II, 369f. See also the neutrino mass web page at http://physics.princeton.edu/borexino/nu-mass.html.

54. The nondynamical character of lepton and baryon numbers is another reason for doubting that such conservation "laws" are correct: Aitchison and Hey 2004, I, 56, 336. See also B. P. Roe, *Particle Physics at the New Millennium* (New York: Springer, 1996), 354, and F. Close, *Antimatter* (New York: Oxford University Press, 2009), 124f. Apart from their lepton numbers, charged leptons differ only in mass and stability; they too may be varieties within one natural kind.

55. Griffiths 2008, 338–46. Also Weinberg 1992, 144–46; and Smolin 1997, 54–57, 72.

56. Griffiths 2008, 375–78, 402. Also A. Salam, *Unification of Fundamental Forces* (New York: Cambridge University Press, 1990), 47–50; A. Zee, *Fearful Symmetry: A Search for Beauty in Modern Physics* (Princeton, NJ: Princeton University Press, 1999), 227–29. That spontaneous symmetry-breaking has occurred is an additional premise; it cannot be derived within the electroweak theory itself.

57. Cf. Weinberg 1992, 144f; L. Smolin 1997, 51–7, 72; A. Zee, *Fearful Symmetry*, 270. Also Auyang 1995, 172.

58. It is remarkable that Weinberg notes the "Platonic" character of this picture without being repelled by it; Weinberg 1992, 195.

59. For more on cases like these, cf. J. Allday *Quarks, Leptons, and the Big Bang Theory*, 53–6, 148–52 198–204, 209; Griffiths 2008, 62, 76, 215f, 246–49, 256, 266, 275 , 327, 331, 334; Auyang 1995, 160–65; E. A. Cornell and C. F. Wieman, "The Bose-Einstein Condensate," *Scientific American* 278 (1998): 40–5.

60. Arguments for this premise can be found in Strawson 1959, 16ff ; French 2015; P. Teller 1990, 606f; and French and Krause 2006.

61. Aitchison and Hey 2003, I, 113–18, 131f, 141–43, 193. Also F. Dyson, "Field Theory," *Scientific American* 188 (1953): 61f; A. Zee, *Fearful Symmetry*, 192f; F. Wilczek, *The Lightness of Being* (New York: Basic Books, 2008), 33, 47, 73, 89–91, 101; Weinberg 2004, 242, and 1995, 2032–40; Auyang 1995, 4f, 115f, 122, 129, 159ff; Casey 2013, 57–88.

62. Aitchison 1985, 334–391. Also T. U. Cao, "Introduction: Conceptual Issues in Quantum Field Theory," and B. DeWitt, "Quantum Field Theory and Space-Time—Formalism and Reality," both in Cao 2004, 9f, 178.

63. Weinberg 1977, 21f. Like Aitchison in the article just cited, Weinberg makes quantum fields the primary entities. Unlike Aitchison, he makes field quanta "epiphenomenal," thereby emphasizing the holistic character of his field-theoretical ontology.

64. For a good introduction to the quantum field, cf. Aitchison and Hey 2003, I, chapters 5–7. For a Leibnizian attempt to capture the real multiplicity of a quantum field without failing to account for its unity, cf. Auyang 1995, chapters 5–7.

65. On the difference between continuants and perdurants, see chapter 4 of this volume.

66. For additional examples, cf. Griffiths 2008, 138; A. Salem, *Unification of Fundamental Forces*, 37f; R. N. Cahn, "The Eighteen Arbitrary Parameters of the Standard Model in Your Everyday Life," *Reviews of Modern Physics* 68 (1996): 951–59.

67. Griffiths 2008, 414–17; F. Wilczek, *The Lightness of Being*, 195–97, 222; D. B. Cline, "The Search for Dark Matter," *Scientific American* 288 (2003): 50–9; C. Seife, "What Is the Universe Made of?" *Science* 309 (2005): 78.

68. e.g., F. Wilczek, *The Lightness of Being*, especially Part III.

69. Griffiths 2008, 52, 84f, 405–13; G. Kane, *Supersymmetry: Unveiling the Ultimate Laws of Nature* (Cambridge, MA: Perseus, 2000); F. Wilczek, *The Lightness of Being*, 177ff; A. Zee, *Fearful Symmetry*, 229–244, 271, 283. Supersymmetry in GUTs doubles the number of fermionic kinds; but these too should not be regarded as natural kinds in the primary sense, since their members are but field excitations and the fields themselves are perdurants rather than continuants.

70. Greene 1999, especially chapter 6. I follow Greene in focusing on strings that are closed.

71. Ten spatial dimensions are posited in what E. Witten and others call "M theory." The additional objects are called "2–branes," "3–branes," etc. Cf. Greene 1999, chapter 12.

72. See chapters 3, 4, and 5 of this volume.

73. A Guth, *The Inflationary Universe*, 86, 136–144, 334, 343. Also J. Allday, *Quarks, Leptons and the Big Bang*, 231, 264; B. Greene 1999, 217, 223; J. Silk, *The Big Bang* (New York: W. H. Freeman, 1989), 113–15; A. Zee, *Fearful Symmetry*, 214f; S. Auyang 1995, 223–28.

74. A. Linde, "The Self-Reproducing Universe," *Scientific American* 281 (November 1994): 48–55; M. Tegmark, "Parallel Universes," *Scientific American* 290 (2003): 41–51: R. Bousso and J. Polchinski, "The String Theory Landscape," *Scientific American* (2004): 60–9; J. D. Barrow and J. K. Webb, "Inconstant Constants" *Scientific American* 292 (2005): 56–63; J. K. Webb, "Are the Laws of Nature Changing with Time?" www.phys.unsw.edu.au/astro/research/PWAPR03webb.pdf; L. Smolin, *Time Reborn* (Boston: Houghton Mifflin Harcourt, 2013), xxxi, 13–24, 95, 118f, 172–91, 208, 240–51. Also A. Guth, *The Inflationary Universe*, 247–51, B. Greene, *The Elegant Universe*, 366–69, and M. Fisk, *Nature and Necessity* (Bloomington, IN: Indiana University Press, 1973), 197–228.

75. L. Smolin in, e.g., Smolin 1997, especially Parts 2 and 3, pp. 105–08, 146–48. Compare J. Silk, "Holistic Cosmology," *Science* 277 (1997): 644.

76. Cf. Smolin 1997, 94 with 15–8, 77; and his *Time Reborn*, 104, 123–39. He assumes that the asymmetry of time is ontologically basic and not a derivative appearance. Whether this assumption is true is currently a matter of debate.

77. E.g., Cassirer 1956, 131f, 178ff. Also Russell 1912, 180–208; *The Analysis of Matter* (London: Routledge & Kegan Paul, 1927); and *Our Knowledge of the External World*

(London: Routledge, 1993), 42, 48. Also Quine 1976; Teller 1986, 71–81; French 1998, 93–113; Chakravartty 2000, S867–S878; Belousek 2003, 791–811; Ladyman and Ross 2007.

78. Cf. M. Durato, "Substantivalism, Relationism, and Spacetime Structuralism," *Foundations of Physics* 30 (2000): 1605–28. Physicists are interested in transformations that do not change the *action*. Perhaps, then, we should think of the world as energy "at work," distributed and redistributed in spacetime, and not as a substance enduring in time.

79. Cf. A. Pais, *Niels Bohr's Times, in Physics, Philosophy, and Polity* (New York: Oxford University Press, 1991), 312–16, 424–47.

80. Mathematicians purified analysis by removing from it all talk of variables; cf., e.g., G. Frege, "What is a Function," in *Translations from the Philosophical Writings of Gottlob Frege*, ed. P. Geach and M. Black (Oxford: Blackwell, 1970), 107–16. For an account of functions that stays closer to the origins of analysis in classical physics, cf. M. Comenetz, *Calculus: The Elements* (Singapore: World Scientific, 2002), 28–35.

81. The term "g," regarded here as a constant, has the dimensions of acceleration (in the cgs system, centimeters per second per second, or $c\text{-}s^{-2}$). So when $t$ = 2 sec, $S$ = 2g cm.

82. Seventeenth-century physicists sometimes expressed laws of nature as proportions rather than equations. So, for instance, instead of "S = ½gt2" they would write "$S_1{:}S_2{::}t_1^2{:}t_2^2$." Descartes' introduction of real numbers, to be used as a sort of common currency, allowed them to alternate the second and third terms: "$S_1{:}t_1^2{::}S_{2?}{:}t_2^2$." Either way, the two ratios are the same. A ratio (*ratio, logos*) is a *relation* with respect to amount between two quantities of the same sort (Euclid, *Elements* V Def 3). In the case of bodies in freefall near the surface of the earth, Galileo found that no matter how the times (or distances) varied, the ratio remained invariant. Such invariances he and other scientists of the time called "laws."

83. Cf. R. Feynman, "The Principle of Least Action," in *The Feynman Lectures*, II-19; Auyang 1995, 64, 68f, 183–87. Also R. Feynman, *The Character of Physical Law* (Cambridge, MA: MIT Press, 1965), 50–53, and A. Zee, *Fearful Symmetry*, 111.

84. That the notion of invariance is modal, see J. Woodward, "Realism about Laws," *Erkenntnis* 36 (1992): 202, and "Explanation and Invariance in the Special Sciences," *British Journal for the Philosophy of Science* 51 (2000): 205, 235; also E. P. Wigner, "The Role of Invariance Principles in Natural Philosophy," in Wigner 1967, 29.

85. Resemblance relations are also internal, and merely ordinal relations may also be necessary (e.g., the darker-than relation between crimson and pink).

86. Cf. J. Earman and J. Roberts, "*Ceteris Paribus*, There Is No Problem of Provisos," *Synthese* 118 (1999): 439–78; J. Woodward, "There is No Such Thing as a *Ceteris Paribus* law," *Erkenntnis* 57 (2002): 303–28.

87. Cf. Bird 2007, 208f.

88. That nomic universals are extrinsic, second-order relations: e.g., Armstrong 1985. That they are internal relations: e.g., Swoyer 1982, 203–23; Bird 2007. Unlike Armstrong, Bird allows for uninstantiated nomic universals (64, 89), as does M. Tooley in "The Nature of Natural Laws," *Canadian Journal of Philosophy* 74 (1977): 671. T. Maudlin takes nomic relations to be primitive, not analyzable into universals: *The Metaphysics Within Physics* (New York: Oxford University Press, 2007), 17f.

89. Mumford 2004, especially 74–6, 144–49, 193, 199. For Bird's response, cf. Bird 2007, 119–22, 189–203.

90. J. A. Wheeler, "On Recognizing 'Law Without Law,'," *American Journal of Physics* 51 (1983): 398–404; B. van Fraassen, *The Scientific Image* (New York: Oxford University Press, 1980).

91. Cf. E. P. Wigner, "Symmetry and Conservation Laws" and "The Role of Invariance Principles in Natural Philosophy," both in Wigner 1967; A. Zee, *Fearful Symmetry*, 114–19; Auyang 1995, 35; D. J. Gross, "The Role of Symmetry in Fundamental Physics," *Proceedings of the National Academy of Sciences USA* 93 (1996): 14256–59; M. Lange, *Laws and Lawmakers* (New York: Oxford University Press, 2009), 104–19.

92. I just now distinguished four possible answers to the question, "Do LNs exist?" In chapter 5, I distinguished four possible answers to the question, "Do NKs exist?" Since answers to one question are logically independent of answers to the other, there are sixteen possible answers to the question, "Do NKs and LNs exist?" Of these, four affirm realism with respect to both. Scientific essentialists today tend to be unqualified NK-realists *and* unqualified LN-realists.

93. Lowe 2006, 93f, 127f, 132, 157f, 162. Armstrong, a fact-ontologist rather than a substance-ontologist, encounters a similar difficulty when he tries to extend his account of second-order nomic relations between properties (or state-of-affairs types) to functional laws; D. M. Armstrong 1985, 101, 111–16.

94. Ellis 2001, 2, 9, 18f, 25, 48, 72, 97–9, 109, 109–35; and his 2002, 1–4, 32, 46, 59, 60–3, 77.

95. Ellis 2001, 25, 53, 97–9, 225f, 276, and his 2002, 4f, 12f, 17f, 28, 56–8, 67, 111–18.

96. Ellis 2001, 12f, 67, 231–34, 255f, and his 2002, 2, 111ff.

97. Ellis 2001, 61–65, 206f, 279, 292, and his 2002, 16, 26–8, 33, 50, 67, 82, 86, 109–15, 139.

98. Ellis 2001, 206f, 265, 267f (compare 92), and his 2002, 36, 13, 87. It is both fitting and odd, then, that he says nothing about the role of photons in Coulombic interactions. In fact he says nothing about quantum *interactions*. In this respect, his account of Coulomb's Law resembles Huygen's treatise on impact laws.

99. Ellis 2001, 51f, 73, 104n22, 172, 205, 268; 2002, 69f, 85–8, 171–76; and 2005, 462–72.

100. Ellis 2001, 172, 204, 224, 249–53, 272–78, and his 2002, 87, 118–20. See also Bigelow, Ellis, and Lierse 1992, 371–88.

101. Cf. Cartwright 1983, 91–100, and *The Dappled World* (New York: Cambridge University Press, 1999).

# SEVEN

# Biological Kinds

Whereas I am examining scientific theories with a view to determining whether natural-kinds realism is true, most essentialists today *assume* that it is true and use scientific theories to determine what natural kinds or their members are essentially. So, for example, scientific essentialists look to chemistry or physics to say what cesium or an electron is; and pre-scientific essentialists look to chemistry or biology to say what water or a tiger is. In chapter 6 we found some justification for the belief that cesium, understood as a class of s-resembling free molecules or atoms, is a generic if not specific natural kind; and some justification for the belief that the class of electrons is not a natural kind in the primary sense, since its members are not individual continuants. Let us now see whether biological theory provides any justification for the everyday assumption, accepted by Putnam and Kripke among others, that *the tiger* is a natural kind whose members are s-resembling continuants.

## SPECIES

The foremost natural kinds are commonsensically thought to be of animals and plants. Aristotelians agreed. In their view, a species is a class of animals or plants whose members exemplify the same substantive kind in virtue of their intrinsic forms. Such specimens were their exemplary stereotokens, and real definitions of the species to which these stereotokens belong were among the stereotypes they sought to articulate. That members of a species are the same essentially, in virtue of their intrinsic forms, does not guarantee that they are the same in all salient respects. Aristotelian species allow for many individual differences, some of which may be regarded as deviations from the norm implicit in the stereotype. For example, some humans may fall a little short of being what it is to be a human being, in which case they are not good specimens of humankind.

Darwinians disagree. In their view, a species is a group of living things whose members are all descended from a common ancestor. That living things resemble one another closely does not guarantee that they belong to the same natural group; there are in fact many cases of analogy (homoplasy) without common ancestry (homology). Conversely, that members of a species constitute a sort of family does not guarantee that they resemble one another closely; they are likely, in fact, to become increasingly dissimilar as heritable differences accumulate. Darwinian species may come into being and perish through heritable variation, differential fitness, and natural selection. Differential fitness does not entail the existence of natural norms. It may be that some humans are less fit than others, but they do not therefore fall short of what it is to be a human being. There are no good or bad specimens of *Homo sapiens*, and no such thing as human nature.[1]

The Darwinian conception of species prevails in biological theory. In practice, of course, biologists may employ the Aristotelian or everyday conception as well—especially when getting their bearings as naturalists in the field or as taxonomists in pursuit of the correct phylogenetic system—and this, together with the fact that the Aristotelian conception remains deeply ingrained in our thinking, is a major source of confusion. But the ultimate source may be everyday thought, according to which exemplary natural kinds are also kinship groups. Perhaps the words "tiger" and "oak" have always designated Darwinian as well as Aristotelian species. To avoid any ensuing confusion, let us distinguish between typological species ($species_t$) and genealogical species ($species_g$).[2] Humankind is a $species_t$, *Homo sapiens* a $species_g$. The former is an extensionally variable class whose members (tokens) s-resemble one another and instantiate one and the same type. Resemblance and instantiation relations are ahistorical.[3] Humankind, considered either as a type or as a class of s-resembling tokens, is not spatially or temporally restricted. It would not necessarily become extinct once and for all if the number of existing human beings went to zero. *Homo sapiens*, on the other hand, is a mereologically variable group whose members (parts) are descended from one and the same ancestor. This ancestor is a particular individual or group, not a type or even a prototype. And the ancestral-descendant relation wherein every member of the group stands to it is plainly historical.[4] Thus *Homo sapiens* is spatially and temporally restricted. It will become extinct once and for all when the number of humans goes to zero.

$Species_g$ are particulars, not universals or their extensions.[5] What sort of particulars are they? Certainly not bare or thin, and their involvement in the world is certainly not grade I. They must then be thick particulars whose involvement in the world is grade II or multitudes of particulars whose collective involvement is grade III. Or are they to be regarded not as endurants but as four-dimensional perdurants? Evolutionary biologists do not distinguish clearly between these two conceptions. On the one hand they regard $\text{species}_g$ as lineages, things essentially temporal as well as spatial, and represent them by the line segments or tubes of a dendrogram.[6] So conceived, they are but parts of a frozen stream of life.[7] On the other hand biologists regard $\text{species}_g$ as the fundamental units of evolution, things enduring in time and eventually ceasing to be as other species come to be out of them. So conceived, they are actually evolving continuant-like things, the many current streamings of life on Earth.

Let $\text{species}_g$ be continuant-like particulars. How like continuants are they? Not very, it seems, for so caught up are they in evolutionary processes that there appears to be no fact of the matter enabling us to determine when any of them begins or ends. Contrary to what the line segments in a dendrogram suggest, there is nothing point-like about the origin or extinction of such a particular, and nothing about its spatial extent that is not ineliminably vague. $\text{Species}_g$ are thus like plumes of smoke. Their principal parts, moreover, appear in many if not all cases to be individuals; and again, no individual can be composed of actual individuals. Thus, compositionally as well as metabolically and situationally, $\text{species}_g$ are grade III and not grade II particulars. It would be a mistake to classify them as continuants.

Higher-order taxa are even less continuant-like. With regard to all biological taxa, then, antirealism seems right. A taxonomist's allocation of living things to clearly identified and distinguished kinship groups may have some basis in evolutionary fact, and insofar as it does we may regard those groups as real; yet as clearly identifiable and distinguishable, they are *pragmata* stipulated and not realities discovered. When it comes to higher-order taxa, many biologists and philosophers of biology accept this conclusion.[8] Yet when it comes to species, many disagree. There is something special about species, they think, which makes them real unities in a way that other taxa are not—something in virtue of which they remain the same as they change and are countably diverse—as are individual continuants. What could this "something" be? It cannot be merely diagnostic.[9] Nor can it be the common-ancestry condition alone, which all biological taxa meet. So there must be an additional condition, probably trans-historical, which species alone meet. What could it be?

To this question species-realists have given different answers depending on which "species concept" they accept. These include the following: (1) the biological species concept, according to which a species is a group of interbreeding, reproductively isolated organisms;[10] (2) the recognition species concept, according to which a species is the most inclusive group of biparental organisms sharing common mate recognition and fertilization systems;[11] (3) the cohesion species concept, according to which a species is the most inclusive group of organisms capable of genotypic and phenotypic commonality through developmental and other intrinsic constraints;[12] (4) the ecological species concept, according to which a species is a group of organisms occupying an adaptive zone minimally different from any other group in its range and evolving separately from all populations outside its range.[13] These concepts yield different taxonomies. For instance, according to (1) no clonal populations is a species, whereas some of them are according to (2)–(4). In biology today, species-realists disagree about what it is that makes a species$_g$ a real unity. Their disagreement seems real, and very difficult to resolve.[14]

We wonder, therefore, whether any species-realist knows what a species$_g$ is, and whether species-antirealism is not after all the stronger position.[15] The question is not whether species are individual continuants—they are not, since they have no inner natures or essences—but whether each species is sufficiently continuant-like that it would be illuminating, not merely convenient, to classify it as an enduring singular thing. To answer this question it helps to give ourselves an ordered series of continuant-like particulars ranging from paradigmatic to marginal. I suggest the following: squirrels, jellyfish, aspen groves, ant colonies, hurricanes, ecosystems, planetary systems, galaxies, clusters of galaxies. Where on this rough ordinal scale do species$_g$ fall? Different concepts pick out different integration and isolation factors to explain why things of this sort are distinct unities, yet none of them makes an entire species of ant as continuant-like as an ant colony is, or *Homo sapiens* as continuant-like as Hurricane Katrina was. Perhaps species fall somewhere a little to the left of ecosystems. But then how illuminating is the analogy between these "units of evolution" and those "units of selection" which are paradigmatic individuals?

Be that as it may, for many evolutionary theorists today species are *stereotokens* whose corresponding *stereotype* their species-concepts are meant to declare. Insofar as there is disagreement about the adequacy of these concepts, the stereotype remains unsettled, yet that species are tokens of the type species-hood is presupposed. Thus has typological thinking persisted within a phylogenetic framework. One might even argue that species$_g$ constitute a natural kind in some secondary sense of the term.[16]

One more thing. The existence of biotic species$_g$ does not preclude the existence of biotic species$_t$. They may even coincide. Humans, for instance, may be at once parts of *Homo sapiens* and specimens of humankind. Biologists are little inclined to consider this possibility. Indeed, they have reason to doubt that humans are the individuals they appear to be, as we shall presently see.

## CODONS

Whereas evolutionary theory remains predominantly historical, genetics has been typological from the outset. Mendel posited heritable factors, later called "genes," which he took to be responsible for the various traits of a living thing or its parts. Although traits often blend into one another, so that one can hardly tell them apart, the genes themselves are discrete both as tokens and as types. The task is to discover these genotypic factors and explain how they give rise to that phenotypic ensemble. Geneticists located the genes first in chromosomes generally, then in specific macromolecules (DNA, RNA). They also began to investigate how these molecular genes give rise to macromolecules of another specific sort (proteins).[17] Proteins and nucleic acids are polymeric. The basic parts of proteins are discrete amino acids. The corresponding parts of molecular genes, likewise discrete, are called "codons."

Each codon is composed of three nucleotides held together mainly by the ribose-phosphate "backbone" to which they are bound. In RNA molecules, these nucleotides are of four specific kinds: adenine (A), cytosine (C), guanine (G), and uracil (U).[18] Hence there are sixty-four possible codon types—AAA through UUU—all realized in the sense that tokens of them are present in terrestrial life. These tokens encode either a specific amino acid or a "stop": AAA codes for lysine, UUU for phenylalanine; UAA, UAG, and UGA are stop codons; and so forth. Conversely, of the many possible amino-acid types, twenty occur naturally in the sense that tokens of them are present in terrestrial life, and to every one of these there corresponds one or more codon types: alanine is encoded by GCU, GCC, GCA, and GCG; histidine, by CAU and CAC; methionine, by AUG; and so forth. All this can be set out neatly in two tables, each the inverse of the other.[19]

When set out in tabular form, the genetic code resembles the periodic table of elements and the table of elementary particles (see Table 6.1). Like them, it is typological, not historical; and like Table 6.1, in particular, its disjoint places are all filled. Yet there are also dissimilarities. For one thing, there is nothing periodic or generational about the genetic code. Nor does it make any claim to universality: life elsewhere in the universe may depend on very different codes.[20] Nor is it exceptionless: UGA codes instead for tryptophan in the nuclei of *Mycoplasma* and in all mitochondria except those of plants; AAA codes instead for asparagine in the mitochondria of flatworms and echinoderms; and so forth.[21] But the greatest difference is that the genetic code is a *code*. In tabular form it constitutes a sort of nucleic-acid/amino-acid lexicon; it tells us what the various nucleotide triplets "mean." Most of them "translate" into one or another specific amino acid. Some are even "synonymous" in the sense that they translate into the same amino-acid type. Codons are *functional* units.[22] There is nothing quite like this in any of the physical sciences, and the difference appears fundamental.

What is a codon? Chemically, it resembles three conjoined molecules *x*, y, and *z*, each of them a hylo-morphic entity of the sort described in chapter 6. As such, a codon token is compositionally grade III and not an entity in its own right. Biologically, however, *xyz* is a single "word" in virtue of its function. As Max Delbrück observed, the conception here is basically Aristotelian: the nucleotides *x*, *y*, and *z* are the material cause of a codon, like the three letters in the word "cat," while its proper function (*ergon*) is the intrinsic formal cause, like that which makes "cat" a single word and not just three letters.[23] Codons are hylo-ergonic particulars present in all terrestrial life.

As two molecular tokens are the same in kind just in case they are both iso-morphic and iso-meric, so also two codons are the same in kind just in case they are both iso-ergonic and iso-meric. Though hetero-meric, tokens of AAA and AAG are iso-ergonic since they all code for lysine. Geneticists regard them as synonymous rather than identical in kind.[24] Though iso-meric, tokens of AUA in the mitochondria of yeast and tokens of AUA elsewhere in living things are hetero-ergonic, since in the former they code for isoleucine, whereas in the latter they code for methionine: they are heteronymous. Geneticists put them in separate tables, one for "the genetic code," the other for exceptions to that code. Together these two tables set out all the codon types. Molecular-functional (biochemical) descriptions of these types are among the principal stereotypes in genetics today.[25]

Again we seem to have stumbled upon an essentialist's dream come true, not in evolutionary theory, as we may have wished, but in molecular genetics. Those natural-kinds realists who look to biology for guidance may find in this theory what they were looking for: real stereotokens and stereotypes. But again there are questions: Is it true that these tokens are continuants? How are biotic functions to be identified? Are biotic functions real? These questions I take up in reverse order.

### *The Reality of Biotic Functions*

An eagle's eyes are for seeing, its wings for flying. These are the functions proper to those bodily parts. If any of those parts should change in a way such that the eagle can fly or see only poorly or not at all, this change would be a deviation from the norm implicit in the function proper to that part. With biotic function comes the possibility of malfunction: an eagle's wing can be defective, a rivulet of water cannot; the eagle can be blind, water cannot. We also believe that such functions are there in the living things to which they belong, quite apart from whatever we think about them. We are receptive therefore to the following scientific claims: Hearts are there in our bodies to pump blood. The blood contains hemoglobin, whose function is to carry oxygen to the muscles. If it carries carbon monoxide instead, or if one has sickle cell anemia, the hemoglobin molecules are not functioning as they should. Creutzfeldt-Jakob disease is caused by a proliferation of misfolded proteins called "prions." mRNA molecules convey genetic messages to the sites where amino acids are produced, proteins formed, and in this process things sometimes go awry despite the presence of mechanisms whose functions are to minimize such errors and correct them when they occur.

Yet there are reasons for rejecting realism with respect to biotic functions. First, it rests on a dubious analogy. The concept of a function has its native home in human practices. Spoons, clocks, governmental agencies, the words I am now using: all have functions that can be identified and explained by reference to the intentions of their designers or users. A spoon, for instance, is supposed to work in a certain way; that is its job, what it is there for, its purpose. It may serve this purpose well, badly, or not at all: there is a generally accepted norm by which to evaluate its performance as the utensil it is. In the physical world there are no designers or users, no intentions or purposes.[26] Therefore, the concept of a function has no real application to physical things or their parts.

Second, the exemplary natural science is physics and physical theory does not admit natural norms. It has no room for the idea that some changes of state are cases of failure or success, proper or improper functioning. True, physicists may find it convenient to regard some things as triggers, accelerators, or targets; and chemists speak of catalysts, buffering agents, and the like. But these ways of conceiving their subject matter are broadly pragmatic; they belong to physical science as a human practice involving generally accepted norms according to which things scientific may be judged failures or successes. One should not read this practice into the world itself. In reality, biotic functions are just dispositional properties of bodily parts we deem organic.

Third, one may argue that livings things are so complex that we, given our mental capacities and limitations, must regard some of their subsystems as having functions of the sort in question.[27] As Hume might have said, we humans are so constituted that we naturally expect what we have singled out as eyes, hearts, proteins, mRNA, and the like to continue behaving as they have. When they do, we think they are functioning as they should; and if they do not, if our expectations are disappointed, we think some malfunction has occurred. The norms are in us, not in the dispositions themselves. That we can hardly avoid reading them into those dispositions is not a reason for regarding them as real.[28]

Since this Humean explanation does not argue for antirealism with respect to biotic functions but presupposes it, we may set it aside for now. Two arguments remain. The first draws attention to a questionable analogy. The deployment of analogies is indispensible in natural science. Classical mechanics regarded natural things as machine-like even though machines are artifacts designed, fabricated, and used. Darwin likened natural selection to artificial selection even though in nature the "selective forces" have no ends in view. And genetics describes molecular genes and their components in semantic terms even though such things are without intentionality.[29] Is it the case that these analogies, though fruitful, are without ontological import? Geneticists assert that the RNA nucleotide triplet CAU is word-like in that it *codes for* the amino acid histidine. Is this analogy not illuminating? That remains to be shown.[30]

The second reason draws attention to the fact that the physical sciences do not depend on analogies of the sort peculiar to the biological sciences. What are we to make of this fact? One option is to try reducing the subject matter of molecular genetics to that of molecular chemistry, thereby removing from one's ontology all peculiarly biotic functions. Then the dispositional properties of so-called mRNA molecules in so-called living things would no longer seem any different in kind from the dispositional properties of these same molecules in lifeless physical circumstances. But then the distinction between living things and lifeless homeodynamic systems would also disappear and this distinction, though far from clear, is fundamental for biology—just as the equally unclear distinction between molecules and submolecular quantum ensembles is fundamental for chemistry. In each case there is a liminal condition in which one may witness or at any rate speculate about the emergence of one sort of thing out of another, apparently quite different sort of thing; from a world in which there are no things of a certain kind to a world in which there are things of that kind. Scientists have been busy investigating the transition from a pre-biotic chemical "soup" to living things.[31] Their investigations have been hampered by the fact that concepts proper to the relevant sciences fall short of being what is needed to understand the very emergence of the biotic out of the abiotic.[32]

It does not follow that the subject matter of biology is irreducible to that of chemistry or physics. Yet the distinction between biotic functions and abiotic capacities, as between living and lifeless things, appears fundamental. And since we are using scientific theories to investigate natural kinds, it would be inappropriate for us to ride roughshod over this appearance by maintaining that naturalism generally is reducible to physicalism. We take molecular biology as it is: a well established natural science whose ontological commitments cannot be comprehended by the physical sciences today. We tentatively accept, as twin hypotheses, that living things exist and that some of their parts are functional in the sense described.

### *How are Functional Units to be Identified?*

Two answers have been proposed: the etiological-evolutionary theory[33] and the causal-systemic theory.[34] They agree that such units are related essentially to something beyond themselves, but view the relation very differently. Etiologists conceive it historically: tokens of something having function F must have descended from ancestral tokens naturally selected for F. The proper identification of functional units thus requires phylogenetic research.[35] Systematists, on the other hand, conceive the relation ahistorically: tokens of something having function F must be parts of systems to which they now belong and to whose dynamic stability they contribute by operating as they do. The proper identification of functional units thus requires systemic studies in which phylogenetic research may or may not play a part.

The etiological theory is problematic. First, by appealing to differential fitness and natural selection alone it can hardly explain the normative character of biotic functions. Second, it has trouble dealing with cases of exaptation, where parts originally selected for one function now have another very different function—birds' feathers, for example, or insects' wings, whose main purpose was originally thermoregulation.[36] In such cases it is *as if* the intentions of the original designers are less determinative than those of current users. "Living" languages present many analogous cases. Third, the etiological theory does not accord with biological practice. That eyes are usually for seeing, wings for flying, mitochondria for energy conversion and cellular regulation, nucleotide triplets for encoding amino acids: all this was discovered without any appeal to evolutionary history. True, one's understanding is enriched by finding out how ancestral tokens came to have these functions and why descendant tokens continue to have them; one discovers, for example, that mammalian eyes are not homologous with insect or squid eyes, and finds it reasonable to hypothesize that codons having more than three elements would have been weeded out in early evolutionary history.[37] Similarly, one's understanding of the chemical elements or elementary particles is enriched by investigating how they "evolved." But knowing that history was not needed to *identify* the dispositional properties of matter and antimatter or the isotopes of hydrogen. Similarly, knowing the relevant evolutionary history was not needed to find that cat's eyes and squid's eyes are iso-ergonic albeit hetero-morphic, and thus "synonymous" rather than identical in kind.[38] Nor must geneticists know that codons of some specific type are homologous, not merely analogous, to determine what their specific function is. Probably they are *not* all homologous; and even if they are, biologists will almost certainly never discover that fact.[39/40]

The causal-systemic theory has difficulties of its own. It too has trouble explaining the normative character of biotic functions. Moreover, it describes hierarchically ordered complex systems in general, not living organisms in particular. An amoeba is a complex system whose parts contribute to the activity of the whole. But so is a hydrogen atom; its parts, too, contribute to the atom as a whole, yet their capacities are not functions in the relevant sense.[41] What distinguishes biotic from abiotic systems? No one seems to know. In any case, when applied to living organisms the causal-systemic theory can account for the way in which biologists identify functional units. But there are complications; a given part may have several functions or no function at all, or it may not have the function it appears to have. To settle the matter, one must determine the role that parts of some sort plays in the lives of the organisms to which they belongs. As one textbook puts it:

> There is . . . no simple recipe for discovering a [molecular] gene's function, and no simple standard universal format for describing it. We may discover, for example, that the product of a given gene catalyzes a certain chemical reaction, and yet have no idea how or why that reaction is important to the organism. The functional characterization of each new family of gene products, unlike the description of the gene sequences, presents a fresh challenge to the biologist's ingenuity. Moreover, the function of a gene is never fully understood until we learn its role in the life of the organism as a whole. To make ultimate sense of gene functions, therefore, we have to study whole organisms, not just molecules or cells.[42]

In studies of this sort, evolutionary history is often irrelevant. To find out whether or how a bird's feathers facilitate flight, it hardly helps to know that for its remote ancestors they probably facilitated thermoregulation instead. And to find out whether introns facilitate recombination, it probably does not help to ask whether they are fossilized remnants of dead transposons. Instead, one must concentrate on their possible role in recombination and on the role of recombination in the reproduction of organic wholes.

### *Codons Are Not Continuants*

Supposing biotic functions are real (our first point), and that they are identifiable in a way that agrees with the causal-systemic theory (our second point), it remains to determine the ontic status of codons. In molecular biology, descriptions of them are among the principal stereotypes. But are the corresponding stereotokens really continuants? According to the causal-systemic theory, these hylo-ergonic units are essentially dependent, directly or indirectly, on the organisms to which they belong. Their situational involvement in the world is therefore grade III, not grade II. They would be continuants, perhaps, if they were just molecules, if they belonged entirely to the subject matter of physical chemistry. In fact, however, they have been taken up into the life-activities of organisms operating in their usual environments; they must be considered *bio*chemically, as functional units radically situated in the organisms to which they belong. Molecular biologists tend to regard these functional units as real unities, entities in their own right. Similarly, atomic physicists tend to regard as real unities the subatomic parts of atoms. But just as these parts cannot in actuality be continuants if the atom to which they belong is a continuant, so too the codons and other functional units in an organism cannot be continuants if the organism is a continuant. Organisms are (1) the principal biotic systems to which such units belong as essentially dependent parts, and (2) the principal building blocks of the species$_g$ to which they belong as essentially independent parts. Biologists, it seems, have every reason to regard them as *the* continuants without which their science would have no subject matter. But their theories give one pause, as I now proceed to show.

## LIVING ORGANISMS

Biology shares with other natural sciences a two-fold tendency: to regard macroscopic things as derivative upon their micro-constituents, and to regard things generally as relative to one another in space and time. Thus, in molecular biology, a living thing appears to be but a manifestation of complex molecular processes. In developmental biology, morphogenesis appears to be nothing more than a modular cascade of microprocesses in a mosaic of competing cell lines.[43] In ecology, an ecosystem appears to be a dynamic structure whose subsystems, including living things, are but nodes isolable only in thought from the network of relations which constitute the system as a whole.[44] And in evolutionary theory, living things are submerged in the sweep of history; selective forces acting on heritable differences adapt them to their surroundings and these adaptations occur in piecemeal fashion; the true replicators are the genes, which produce living things as interactive vehicles for the perpetuation of themselves.[45] In all these ways, living things appear to be so deeply involved in the world—metabolically, compositionally, and situationally—that they could not be continuants. This consequence is theoretical. In practice, biologists go on thinking of plants and animals as individuals, just as chemists go on thinking of water as a substance and quantum field theorists go on thinking of electrons as particles. As such, however, these objects are *pragmata*, not entities.

Biologists have yet another reason for giving up realism with respect to individual organisms. In chapter 4, while showing how difficult it is to decide whether Blackie and the like are continuants, I made brief forays into biological science to gather a few "hard cases," and yet my argument there remained under the guidance of common sense.[46] Now that it is under the guidance of biological science, however, we see that such hard cases are the rule and not the exception. Again let us give ourselves an ordinal scale ranging from apparently individual to apparently dividual multicellular life-forms. Toward the individual end we put mammals; toward the dividual end, clonal populations of dandelions and aphids; and in between we put zoological and botanical cases of obligate symbiosis, botanical cases in which cells rather than entire plants appear to be the basic units, etc.[47] Everything mentioned here is a complex dynamical system featuring populations of cells. The only real difference between these systems seems to be that in some cases the interactions among its parts are comparatively strong, while in others they are comparatively weak. Interaction-strength is a matter of degree. At no point along our scale does it appear to be a matter of kind as well. At no point does the relation of part to whole seem to shift from one of comparative independence to one of metaphysical dependence. At no point, then, can we say, "*Now* we have genuine individuals!"

Biological science thus prompts one to reject realism with respect to organisms. But it also prompts one to accept it. For in no other natural science do *self*-terms ("self-organization," "self-regulating"), *auto*-terms ("autocatalyzing," "autoimmunity") and *homeo*-terms ("homeostasis," "homeodynamic") seem indispensable. Moreover, in developmental biology, holism has been preserved in conceptions of the morphogenetic field and in the supposition that morphogenetic processes result usually in the generation of a single life-form.[48] In molecular biology, genes are no longer thought to be isolated entities above the fray. "Bean-bag" genetics is out, and the genocentric view has given way to one in which there is no single molecular command center, no ungoverned governor; every molecular process depends on other molecular processes in the system (organism) they help to sustain.[49] No longer does it seem right to identify the nature of this system with its genotype, or to think that the phenotype (the organism itself) is merely epiphenomenal.[50] In immunology, moreover, the notions of self and nonself, health and disease, are central; and the *self* in the "self-recognition" commonly ascribed to antibodies is properly that of the organism as a whole, including even the eubacteria in its gut tract, and never just to the antibodies themselves.[51] Meanwhile, in ecology it is understood that organisms continue to modify the niches that continue to modify them; they are agents as well as patients.[52] And in evolutionary theory Leo Buss and others have described "the evolution of individuality." Around 500 million years ago, they argue, clonal populations of diversified and mutually attached cell lines became so adapted to one another, the competition between them so constrained, that they became individual organisms whose germ lines were sequestered and whose cellular constituents were now functionally dependent not only on one another but on the organic wholes to which they belonged.[53]

Thus biological science is of two minds about the reality of organic individuals—its analytic bent inclining it to antirealism, its holistic character inclining it to realism. The realism side prevails insofar as it remains rooted in natural history. Yet "organismal thinking" has had a rough go of it. There is today no generally accepted theory of the organism, and no discipline called "organismal biology."[54] Specialization prevails, and with it a tendency to forget, if not reject, the individuality of living organisms. The following passage is representative:

> It is at first sight difficult to accept the idea that each of [these] living creatures is merely a chemical system. The incredible diversity of living forms, their seemingly purposeful behavior, and their ability to grow and reproduce all seem to set them apart from the world of solids, liquids, and gases that chemistry normally describes. Indeed, until the nineteenth century, it was widely accepted that animals contained a Vital Force—an "animus"—that was uniquely responsible for their distinctive properties. We now know there is nothing in living things that disobeys chemical and physical laws.[55]

The authors of this passage suppose (1) that the subject matter of biology is reducible to that of another natural science, in this case chemistry, and (2) that its subject matter would not be so reducible only if living organisms contained some vital force. In other words, these exemplary continuants may be regarded as real only if they contain something that science must regard as unreal.

Both suppositions are questionable. With regard to (1), it is important to distinguish methodological from metaphysical individualism, and methodological from metaphysical "dividualism." Organismal realism takes some if not all living things to be metaphysically individual. Organismal antirealism takes all of them to be metaphysically dividual. One cannot accept both. In biology, however, it is *methodological* dividualism that prevails, and will continue to prevail, in evolutionary theory and ecology as well as molecular and developmental biology. And contrary to what these authors suppose, methodological dividualism does not entail metaphysical dividualism.[56] Therefore, one can be relentlessly analytic in one's study of living organisms and remain an unwavering organismal realist. In other words, one can take such things to be deeply involved in the world—situationally as well as compositionally and metabolically—without taking them to be grade III non-entities. Biologists *can* eat their cake and have it too.[57]

With regard to supposition (2), it suffices to recall our account of continuants in chapter 4. There we found that if Blackie and the like are the entities they appear to be, each is a being-in-becoming, individual in its multiplicity, independent in its situational dependence. In other words, its involvement in the world is grade II. Grade II, we argued, is not resolvable into grade III *plus* grade I, nor can a continuant $x$ be its inner nature $n_x$ *plus* whatever manifold depends causally on $n_x$ for its distinctive properties. But this is just how the Vitalists conceive living organisms. One can refuse to make their mistake without also rejecting organismal realism. Unfortunately, the ghost of Vitalism haunts biologists even today. They could be well rid of it if they were not also (understandably but unjustifiably) leery of metaphysics.

Though biologists are committed to methodological individualism as well as methodological dividualism, one might think they would refrain from making any metaphysical assumptions. But this is false; they do take organismal realism to be acceptable albeit dubious, and organismal realism is a variety of metaphysical individualism. They would be in a better position to defend this thesis if they were less willing to let it be a tacit assumption, and more willing to engage in organismal theory. In any case, living organisms are still among their stereotokens. And we who are now using biological science as a guide have not yet found in it a sufficient reason to reject our earlier conclusion (hypothesis H5) that some of these stereotokens are genuine continuants.

Do biologists recognize any corresponding stereotypes? Perhaps they would count identical twins or quadruplets as instances of s-resemblance classes, perhaps also clonal populations of aphids or dandelions. But we hesitate to regard these as exemplary "natural kinds"; and in any case what we most want to know is whether any $species_g$ (or higher-order taxa) are such that a biologist would be prepared to classify them as $species_t$. In fact, what observant people in different societies regard as $species_t$ of living thing correlates remarkably well with some of what biologists regard as $genera_g$.[58] So the idea that a biological species may also be a natural kind is not initially outlandish. But our inquiry is hampered by the fact that biologists show little interest in what I have called "s-resemblance" or "the nature $nx$ of organism $x$." It is also difficult to fit their ways of classifying organisms into a natural-kinds framework. The phenetic approach treats organisms as clusters of traits without distinguishing between those which are necessary and those which are not.[59] The morphological, functional, and genetic approaches go deeper, it seems, when they focus on basic body-plans, ways of being alive, and genomes. They are also ahistorical. But they do not sort out organisms into the same groups, and may even disagree about which living things are genuine individuals.[60] The phylogenetic approach has given us $species_g$ instead of $species_t$. Being thoroughly historical, its taxonomic results can hardly be brought within a natural-kinds framework. A few philosophers, among them P. E. Griffiths, have tried to do so by proposing natural kinds with historical essences, but that way lies conceptual confusion.[61] Better to begin by noticing once more that nothing rules out the possibility that a $species_g$ be a $species_t$ as well. Though biologists are not much interested in speculative possibilities, perhaps they would consider the following.

(1) Cases of paraphyly. Let there be four populations $p_1 - p_4$ such that $p_3$ is to be classified with $p_4$ but not $p_1 - p_2$ on phylogenetic grounds, but with $p_1 - p_2$ but not $p_4$ on the grounds of interbreeding with reproductive isolation. According to the biological species concept, then, $p_1 - p_3$ are three varieties of a single $species_g$. They may also constitute a single $species_t$. Phylogeneticists cannot reject this possibility without supposing

that the species-concept debate has been resolved in their favor.[62] (2) Cases of hybridization. Suppose two species $s_1$ and $s_2$ hybridize to form a new species $s_3$, whose members reproduce uniparentally. Then $s_3$ is polyphyletic. Now suppose $s_3$ rapidly becomes extinct and soon thereafter a new species $s_4$ originates through further hybridization. Are $s_3$ and $s_4$ not really one and the same recurring $species_t$?[63] (3) Cases of polyploidy. Suppose a species $s_1$ of diploid organisms gives rise to a new species $s_2$ of tetraploidal organisms which compete successfully with their progenitors. Suppose moreover that after some time has passed $s_1$ again gives rise to a species $s_4$ of tetraploidal organisms which interbreed with descendant members of $s_3$. Then $s_3$ and $s_4$ are a single $species_g$ according to the biological species concept, and may be one and the same $species_t$ according to the natural-kinds species concept.[64] (4) Cases of artificial life. Scientists have now implanted an artificial genome in bacterial cells.[65] Suppose they manage to synthesize organisms so similar to the members of a naturally occurring $species_g$ that biologists cannot tell them apart. Are they and their descendants not really members of the same species? (5) A Putnamesque Twin Earth case. Suppose that on a distant planet much like ours there exists a species—call it "*X.y.*"—which so closely resembles *Homo sapiens* that the only major difference between them is phylogenetic.[66] Suppose, moreover, that members of these two $species_g$ come together and interbreed, and that the offspring interbreed with their progenitors. Would not the result of this hybridization with introgression be a single $species_t$, humankind, whose extension was once widely scattered?

Cases (4)–(5) may be so fanciful that biologists will not take them seriously, yet they together with cases (1)–(3) serve to show that there is in biological theory today a possible theoretical role for the *natural-kinds* species concept.

## EMERGENCE

I have on several occasions described the emergence (*genesis*) of a continuant from within a given material situation. What exactly is meant by "emergence" in these contexts?[67]

In fundamental physics, the existing entities are quantum fields (should there be several of them), which are best understood as four-dimensional occurrents. There are also field quanta, indefinitely many of them, but their situational involvement in the fields to which they belong is grade III; they are not continuants. There are also local interactions, indefinitely many of them, but these events are point-like occurrences in spacetime; they do not endure. According to fundamental physics, then, there are no continuants and no natural kinds in the primary sense.

Therefore, if continuants exist, they must be emergent entities; they must have come into being.

To identify which continuants, if any, are admitted in scientific theory, one must turn to the special sciences. Here we found putative continuants (stereotokens) of several sorts: in biology, living organisms; in chemistry, individual molecules and atoms; and in particle physics, perhaps some hadrons, e.g., the quark-composites called "free protons." Living organisms and single molecules, atoms, protons and the like are all thought to be continuants. All of them appear to have emerged within the cellular, molecular, atomic, subatomic, or quantum-theoretical situations on which they depend.

One can order these continuants in the following way: free protons depend directly on field quanta; atoms, on subatomic particles; molecules, on atoms; primitive biota, on abiotic molecules; multicellular organisms, on diverse populations of living cells. We may order the natural sciences accordingly: physics at the bottom, then chemistry, then biology. Natural science thus resembles a layer-cake whose different levels are explained in part by the heterogeneity of the subject matter proper to each layer.[68]

As we descend from ethology to quantum field theory it becomes more and more difficult to argue that the stereotokens we find at some level are genuine individuals: organisms, yes; atoms, maybe; field quanta, no. And as we ascend from fundamental physics to biology, it becomes more and more difficult to argue that the law-like statements we find are expressions of genuine laws of nature in the primary sense of the term: $E = mc^2$, yes; $2H_2O \leftrightarrow H_3O^+ + OH^-$, maybe; Dollo's Law, no. It also becomes more difficult to argue that tokens of nominal thing-types fall neatly into disjoint classes: protons and atoms, yes; macromolecules, perhaps; living organisms, no. Consequently, as exemplary thing-types gradually disappear in the course of our ascent, so too do genuinely nomic statements. But it does not follow that exemplary natural kinds also disappear; for the members of such kinds are exemplary continuants, and these entities appear not to fall neatly into disjoint classes. For natural-kind realists this constitutes a dilemma, one that cannot be resolved by modifying the concept of continuant or the concept of a natural kind.[69]

Working against this layer-cake picture are the reductionist tendencies in natural science. Biologists tend to regard living things as nothing more than systems of cells or molecules; chemists, to regard molecules as nothing more than systems of atoms or subatomic particles; particle physicists, to regard elementary particles as nothing more than normal states of quantum fields. This general tendency and the widespread opposition to it give rise to the following dilemma: either we (1) embrace the idea of a grand theory of everything, according to which so-called emergent entities are continuant-like patterns whose involvement in the world is really grade III, or we (2) embrace the idea that there are genuinely emergent entities, in which case the disunity of natural science appears ineluctable. Each of these opposing views has its advocates. Their debate is ongoing. No one has found a way beyond the opposition they represent.

Since continuants are emergent entities, those of us who agree that such entities exist (hypothesis H5) must also agree that substantial emergence has occurred, that a theory of everything could not be a theory of every *thing*.[70] Yet the notion of substantial emergence remains obscure. We must try to clarify it.

I have argued that the metabolic, compositional, and situational involvement of continuants is grade II. So, if one agrees with B. Ellis that electrons—"real," "effective," "free" electrons[71]—are enduring particulars, one should also agree that these entities have emerged out of electronic field quanta whose situational involvement in the field to which they belong is grade III. And if $H_2$ molecules are enduring particulars, they may have emerged out of the H atoms which are now their constituents. Notice that the emergence of a molecule must have occurred simultaneously with the virtualization or submergence of the atoms as they became parts of it; otherwise the molecule would not be a single entity but several entities. Similarly, when the first multicellular individuals emerged out of diverse populations of associated cells, those cells must have lost their grade II independence as they were taken up in the life of this new whole. Thus, for all continuants except those at the lowest level, emergence and submergence are two sides of one event. Such events are not easily understood. Indeed, the coming into being of a continuant is even more obscure than its being; and its being is quite obscure, as we noted in chapter 4 while trying to understand how something could be really invariant in its real variations, really unitary in its real multiplicity, and really independent of the circumstances on which it really depends.

Scientific theory suggests that new classes of particulars emerge in symmetry-breaking events. This befits the double-sidedness of emergence: continuants of some type "come to light" as a prior state of greater symmetry becomes "hidden."[72] It is unclear, however, whether this suggestion can be generalized to the emergence of each and every continuant. Also unclear is how it might be refined so as to distinguish substantial from nonsubstantial emergence—the emergence of water as distinguished from the emergence of its solid out of its liquid state, or the genesis of a living organism as distinguished from its step-wise maturation. There are scientists working at the "lower" frontiers of their disciplines, trying to understand how life emerged in abiotic circumstances, how molecules emerge in submolecular interactions, etc. But these investigators are hampered by their inattention to the distinction between substantial and nonsubstantial, and by their reductionist tendencies more generally. The concept of *substantial* emergence, required if one is to understand the coming-into-being of those beings-in-becoming we call "continuants," still has no secure place in scientific theory or in current philosophy of science.

It does not follow, however, that contemporary scientific theory cannot help us to adjudicate the debate between natural-kinds realists and antirealists. For although its reductionist tendencies seem to favor antirealism, as Russell and Quine observed,[73] the evident nonreducibility of such continuant-like objects as molecules and living organisms to the objects of quantum field theory lends support to the proposition that there are actually existing continuants (hypothesis H5). And the fact that *some* of these continuants fall neatly into disjoint classes, in virtue of what appear to be their natures or essences, lends support to the proposition that there are also natural kinds in the primary sense of the term (hypothesis H6). So which are the exemplary natural kinds? To judge from our survey of physical and biological theories, they include the class whose members are all and only hydrogen atoms, provided they are isotopic; the class whose members are all and only $H_2O$ molecules, provided they are essentially isomorphic as well as isomeric; and perhaps even a class whose members are polyploidal organisms belonging to the species *Galax urceolata,* or beetleweed, provided they resemble one another in virtue of their natures. These kinds are of little interest to practicing chemists and biologists or taxonomists, and they are not among the kinds regarded as exemplary in everyday thought and speech. Our task, however, was not to show that scientific or prescientific kinds are natural—indeed many of them appear to be pragmatic rather than real—but to determine whether there are in fact any natural kinds in the primary sense of the term. And in this we appear to have succeeded: notwithstanding some residual difficulties, natural-kinds realism appears on reflection to be true.

## NOTES

1. Ghiselin 1997, 76f, and Sober 1980, 350–83.
2. Mayr 1982, 45–47, and 1988, 168–214. See also chapter 5, p. 105 of this volume.
3. In sexual generation, according to Aristotle, an animal of specific kind K begets another animal of the same specific kind. Such recurrences are cyclical, not historical in the Darwinian sense.
4. Since these natural groups (or classes) are mereologically (or extensionally) variable, standard mereology (or set theory) does not apply. No one to my knowledge has tried to axiomatize evolutionary theory within standard mereology, but at least one person, Mary Williams, has tried to axiomatize it within standard set theory: M. B. Williams, "Deducing the Consequences of Evolution," *Journal of Theoretical Biology* 29 (1970): 343–85. Oddly enough, Alexander Rosenberg has defended her approach even though he agrees that species are not sets, and even though he has nowhere shown that the axioms of formal set theory can be so construed as to capture, in any revealing way, the specific particulars in question: Rosenberg 1985, especially 139 with 142. On nonstandard mereology, cf. P. Simons, *Parts: A Study in Ontology* (Oxford: Clarendon Press, 1987).
5. Ghiselin 1997; Hull 1989; Ruse 1987, 225–42.

6. Consider, e.g., G. G. Simpson, *Principles of Animal Taxonomy* (New York: Columbia University Press, 1961), 153; E. O. Wiley, *Phylogenetics: The Theory and Practice of Phylogenetic Systematics* (New York: Wiley-Interscience, 1981), 80; and De Queiroz 1999, 49, 53–54, 65, 82n13.

7. Cf. D. J. Futuyma, *Evolutionary Biology*, 3rd edition (Sunderland, MA: Sinauer, 1998), 92.

8. Cf., e.g., C. Darwin, *On the Origin of Species by Means of Natural Selection* (Cambridge, MA: Harvard University Press, 1964), 52; J. B. S. Haldane, "Can a Species Concept Be Justified?" *Systematics Association* 2 (1926): 95f; G. G. Simpson, *Principles of Animal Taxonomy*, 153; P. H. Raven, "Systematics and Plant Population Biology," *Systematic Biology* 1 (1976): 284–316; E. O. Wiley, "The Evolutionary Species Concept Reconsidered," *Systematic Zoology* 27 (1978): 18; G. Nelson, "Species and Taxa: Systematics and Evolution," in *Speciation and Its Consequences*, ed. D. Otte and J. A. Endler (Sunderland, MA: Sinauer, 1989), 60–81; J. Mallet, "The Speciation Revolution," *Journal of Evolutionary Biology* 14 (2001): 887f; M. Ereshefsky, "Species, Higher Taxa, and the Units of Evolution," in M. Ereshefsky 1992, 381–98; 1998, 103–20; and 1999, 285–305.

9. M. Schilthuizen, "Dualisms and Conflicts in Understanding Speciation," *BioEssays* 22 (2000): 1134–41.

10. E. Mayr, e.g., "Speciation Phenomena in Birds," *American Naturalist* 74 (1940): 256; Mayr 1942, 120; and his "Species, Classification, and Evolution," in *Biodiversity and Evolution*, ed. R. Arai, M. Kato, and Y. Doi (Tokyo: National Science Museum Foundation, 1995), 3–12. M. Ghiselin has argued for one version of it, on explicitly metaphysical grounds, in his 1995, chapters 6–10. J. A. Coyne and H. A. Orr argue for another version in their 2004. That one of Mayr's more recent formulations of the biological species concept limits its primary application to populations at some given time *t*, and may not apply to species as populations evolving over long periods of time: Borgesson 1999,81f.

11. H. E. H. Paterson, "The Recognition Concept of Species," in *Species and Speciation*, ed. E. S. Vrba (Pretoria: Transvaal Museum Monograph 4, 1985), 21–29, reprinted in Ereshefsky 1992, 139–58.

12. A. R. Templeton, "The Meaning of Species and Speciation: A Genetic Perspective," in *Speciation and Its Consequences*, ed. D. Otte and J. A. Endler, 3–27, reprinted in Ereshefsky 1992, 159–86; and J. Mallet, "A Species Definition for the Modern Synthesis," *Trends in Ecology and Evolution* 10 (1997): 294–99. Consider also N. Eldredge and S. J. Gould, "Punctuated Equilibria: An Alternative to Phyletic Gradualism," in *Models in Paleobiology*, ed. T. J. M. Schöpf (San Francisco: Freeman, 1972), 114.

13. L. Van Valen, "Ecological Species, Multispecies, and Oaks," *Taxon* 25 (1976): 233–39, reprinted in Ereshefsky 1992, 69–78. Also S. J. Gould, "The Disparity of the Burgess Shale arthropod fauna and the limits of Cladistic Analysis: Why We Must Strive to Quantify Morphospace," *Paleobiology* 17 (1991): 411–23; Maynard Smith and Szathmáry 1995, 163–67. Several biologists, among them R. Lewontin, have emphasized the conceptual link between a species and its environment.

14. Notice that Darwin, unlike some evolutionary theorists today, did not believe that there is anything metaphysical special about species, except in the minds of those who deny that species, unlike varieties, originate by natural means. In *The Origin of Species*, his preferred term is "race," which he uses indifferently for varieties, species, and genera.

15. For a variety of opinions on this question, see J. Cracraft, "Species Concepts and Speciation Analysis," *Current Ornithology* 1 (1983): 159–87, reprinted in Ereshefsky 1992, 93–120; M. Luckow, "Species Concepts," *Systematic Botany* 20 (1965): 589–605; *Species Concepts and Phylogenetic Theory: A Debate*, ed. Q. D. Wheeler and R. Meier (New York: Columbia University Press, 2000); Coyne and Orr 2004, 25–48, 447–72; J. Mallet, "Species Concepts," http://www.ucl.ac.uk/taxome/jim/pap/mallet05spconc.pdf; T. A. C. Reydon, "Why Does the Species Problem Still Persist?" *BioEssays* 26 (2004): 300–05; D. N. Stamos, *The Species Problem* (Lanham, MD: Lexington Books, 2003). Favoring a single species concept: Hull 1978, 37; 1998, 183, and "The Ontological

Status of Species as Evolutionary Units," in his 1989, 80; and De Queiroz 1998. Favoring pluralism, see B. D. Mishler and M. J. Donoghue, "Species Concepts: A Case for Pluralism," *Systematic Zoology* 31 (1982): 491–503, reprinted in E. Sober 1994; also P. Kitcher, "Species," *Philosophy of Science* 58 (1991): 84–101, reprinted in Ereshefsky 1992, 317–42. That constraints on gene flow do not suffice, see Barker 2007, 654–65; Barker and Wilson 2010, 59–77. On the role of unitary selective processes and developmental or other homeostatic constraints: Hull 1978, 335–60; E. O. Wiley and D. R. Brooks, "Victims of History—A Nonequilibrium Approach to Evolution," *Systematic Zoology* 31 (1982): 1–24. That species may be units of selection, see e.g., R. Lewontin 1970, 1–18. That a species may have its own individual essence, see S. Okasha 2002, 193f.

16. On the question whether higher-order taxa are likewise natural kinds, cf. G. C. D. Griffiths, "The Future of Linnaean Nomenclature," *Systematic Zoology* 25 (1976): 168–73; Ghiselin 1997, 52.

17. The transition from Mendelian to molecular genetics is historically and conceptually complicated; cf. P. Portin, "The Concept of the Gene," *Quarterly Review of Biology* 68 (1993): 173–223.

18. Thymine (T) replaces uracil in a DNA molecule. A and G are purines; C, T, and U are pyrimidines. In a DNA or RNA molecule, pyrimidines are complementary with purines: G with C, A with T or U. In such pairings, the "base pairs" are connected by comparatively weak hydrogen bonds that are easily "unzipped" by the appropriate enzyme. A-C and G-T pairings are unlikely on electrochemical grounds. In RNA molecules, however, the "wobble" base pair G-U pairing is frequent and functionally important. Cf., e.g., A. J. F. Griffiths, J. H. Miller, D. T. Suzuki, R. C. Lewontin, and W. M. Gelbart, *An Introduction to Genetic Analysis*, 7th edition (New York: Freeman, 2000), especially 312–20; B. Alberts et al. 2004, especially 335f.

19. AUG functions also as an initiation codon; GUG and UUG are sometimes initiation codons in bacteria. For examples of such tables, see any introductory molecular genetics textbook.

20. It is doubtful, therefore, that biologists have discovered any genetic laws in the primary sense of the term. Cf. Hull 1978, 335–60.

21. For a list of such exceptions, cf., e.g., R. D. M. Page and E. C. Holmes, *Molecular Evolution: A Phylogenetic Approach* (Oxford: Blackwell Science, 1998), 47. See also Alberts et al 2004, 352f, 434, 814.

22. "Synonymy" is common in functional systems. It is unsurprising that living things have more than one way of producing some amino acids, and wonderful but hardly astonishing that while UGG codes for tryptophan in many cases, in the nuclear genome of *Mycoplasma* it is UGA (usually a stop codon) which codes for this amino acid. Similarly, while it is usually the iron in hemoglobin which picks up free oxygen, in crustaceans it is copper. In functional systems there is often "more than one way to skin a cat."

23. M. Delbrück, "Aris-totle-totle," in *Of Microbes and Life*, ed. J. Monod and E. Borek (New York: Columbia University Press, 1971).

24. Compare G. C. Williams, *Natural Section: Domains, Levels, and Challenges* (New York: Oxford University Press, 1992), 11.

25. The classification of proteins in molecular biology involves similar considerations: M. M. Slater, "Macromolecular Pluralism," *Philosophy of Science* 76 (2009): 851–63.

26. Early modern natural philosophers did sometimes regard God as the designer of things natural. But I do not find that Harvey, in his *Motion of the Heart and Blood,* relied on this theological view when determining that the heart is like a pump. And Aristotle could not have relied on it even tacitly when determining the functions of various animal parts, since he held no such theological view.

27. Kant made an argument along these lines in his *Critique of Judgment*.

28. Davies 2001, 3, 5, 8, 11, 14–15, 75, 87–98, 136–37, 141, 150–53, 175–79, 182, 214f.

29. That molecular biologists could remove all intentional language from their discourse, see Rosenberg 1985, 260f. But the more basic functional language would still remain.

30. For more on this issue, cf. M. Hesse, *Models and Analogies in Science* (Notre Dame, IN: Notre Dame University Press, 1970), and Umphrey 2002, chapter 5.

31. Here, as in other such transitions, symmetry-breaking events are discovered or postulated.

32. Attempts to apply the concept of natural selection beyond the biotic domain are problematic, as are attempts to specify living things within a general theory of complex chemical systems. Some physicalists have tried to erase the biotic-abiotic distinction by recourse to the notion of information; cf., e.g., Ladyman and Ross 2007, 183–89, 210f, 238–57, 279f. But the notion of information is notoriously slippery; for cautionary remarks from a biologist, cf. Maynard Smith 2000, 177–94, esp. 178f, 183–85, and 189–93.

33. For presentations and defenses of this theory, cf. Wright 1973, 139–68, and Millikan 1989, 288–302; Neander 1991, 168–84.

34. For presentations and defenses of this alternative theory, cf. Cummins 1975, 741–60, Kitcher 1993, 379–97; Amundson and Lauder 1993–1994, 443–69; Preston 1998, 215–54; P. W. Reeve and H. K. Sherman, "Adaptation and the Goals of Evolutionary Research," *Quarterly Review of Biology* 68 (1993): 1–32; N. Tinbergen, "On the Aims and Methods of Ethology," *Zeitschrift für Tierpsychologie* 20 (1963): 410–33, and "On War and Peace in Animals and Man," *Science* 160 (1968): 1411–18; Mayr 1961, 1501f.

35. Contrary to what some of its advocates believe, the etiological theory cannot ultimately explain the presence of biotic functions, since it presupposes the existence of things that are alive and things are alive only if they have parts some of whose capacities are already functions.

36. W. J. Bock, "Explanatory History of the Origin of Feathers," and S. S. Sumida, C. A. Brochu, "Phylogenetic Context for the Origin of Feathers," *American Zoologist* 40 (2000): 478–85, 486–50; compare R. O. Prum and A. H. Brush, "Which Came First, the Feather or the Bird?" *Scientific American* 288 (2003): 84–93; J. G. Kingsolver and M. A. R. Koehl, "Aerodynamics, Thermoregulation, and the Origin of Insect Wings: Differential Scaling and Evolutionary Change," *Evolution* 39 (1985): 488–504, and "Selective Factors in the Evolution of Insect Wings," *Annual Review of Entomology* 39 (1994): 425–51.

37. P. V. Baranov, M. Venin, and G. Provan, "Codon Size Reduction as the Origin of the Genetic Code," www.ncbi.nlm.hih.gov/pubmed/19479032. There may also be an evolutionary explanation of the fact synonymous codons such as ACC and ACG usually differ only in the third, "wobble" position: A. Wagner, "Distributed Robustness Versus Redundancy as Causes of Mutational Robustness," *BioEssays* 27 (2005): 176–88.

38. Compare Hull 1989, 222–27, 237.

39. Tokens of a specific codon type have the same function and the same chemical structure, yet given the frequency of genetic mutations (whether naturally occurring or not) and the likelihood that early life on earth was not monophyletic, there is little reason to suppose that these tokens are homologous; cf. W. F. Doolittle, "Uprooting the Tree of Life," *Scientific American* 282 (2000): 90–95. And although systematists insist on distinguishing between homology and homoplasy, it can be difficult to tell them apart. Take for instance the fact that marine animals at the larval stage have the same basic form across phyla. Homology or homoplasy? A little of both? It may be impossible to tell. Cf. P. Willmer, *Invertebrate Relationships: Patterns in Animal Evolution* (New York: Cambridge University Press, 1990), 35. With regard to the early history of prokaryotes, too, the difficulties appear quite insurmountable: Y. Boucher, C. J. Douady, R. T. Papke, et al, "Lateral Gene Transfer and the Origins of Prokaryotic Groups," *Annual Review of Genetics* 37 (2003): 283–328; E. Bapteste and Y. Boucher, "Revising the Concept of Lineage in Prokaryotes: a Phylogenetic Perspective," *BioEssays* 31 (2009): 526–36.

40. Some advocates of the etiological theory seem to think that things biotic can only be known from a phylogenetic perspective. There are indeed efforts to unite ecology and developmental biology with evolutionary theory, but these efforts are integrative rather than reductive. Among evolutionary biologists, moreover, it is not generally held that natural selection must be able to explain, for example, why one rather than another arbitrary genetic code was realized in what appears to be a case of symmetry breaking, or how it came to be that in all terrestrial life one finds only left-handed isomers of amino acids, when right-handed isomers are easily synthesized. Cf. Pauling 1970, 774–76; D. Futuyma, *Evolutionary Biology*, 166; A. J. F. Griffiths et al., *An Introduction to Genetic Analysis*, 315f.

41. H. Simon, "The Organization of Complex Systems," in *Hierarchy Theory: The Challenge of Complex Systems*, ed. H. Pattee (New York: Braziller, 1973), 1–27, esp. 9. A. Rosenberg tries to mitigate this difficulty in his 1985, 58, 114. But while arguing that biotic functions are underwritten physico-chemically, he assumes they have already been identified biologically. Similarly, atomic theory underwrote the periodic table of elements that had already been identified chemically.

42. Alberts et al. 2004, 27.

43. A. Minelli, *Forms of Becoming: The Evolutionary Biology of Development*, tr. M. Epstein (Princeton, NJ: Princeton University Press, 2009), 35–37, 77–93, 110f, 161f, 180f; Gilbert and Epel 2009, 336ff, 413f.

44. C. Elton, *Animal Ecology* (New York: Macmillan, 1927), 63 et passim; Sterelny and Griffiths 1999, 186.

45. F. Jacob, "Evolution and Tinkering," *Science* 196 (1977): 1161–66; R. Dawkins, *The Selfish Gene* (New York: Oxford University Press, 1976); Hull 1980, 318.

46. Chapter 4.3 of this volume.

47. Cf. J. Wilson 1999, chapters 1–3; Hull 1998, 184–85. For details, cf., e.g., E. O. Wilson, *Sociobiology* (Cambridge, MA: Harvard University Press, 1975), 386; F. J. Little, Jr., "Bispecific Chimerical Sponges from the Antarctic," *Nature* 211 (1966): 436–38; D. H. Janzen, "What Are Dandelions and Aphids?" *American Naturalist* 111 (1977): 586–89 with J. Harper, *Population Biology of Plants* (New York: Academic Press, 1977), 27.

48. Cf. C. H. Waddington, "Morphogenetic Fields," *Science Congress* 114 (1934): 336–46; B. C. Goodwin, *How the Leopard Changed Its Spots: The Evolution of Complexity* (New York: Charles Scribner's Sons, 1994), 116–33; G. Webster and B. Goodwin, *Form and Transformation: Generative and Relational Principles in Biology* (New York: Cambridge University Press, 1996), 215–21. Also Alberts et al. 2004, 1242–57; A. Minelli, *Forms of Becoming: The Evolutionary Biology of Development*, tr. M. Epstein, 48, 161, 180f.

49. Cf. Alberts et al. 2004, 67–68, 156–58, 171–72, 186, 214ff, 248, 267–72, 315–19, 338–39, 353–55, 376, 379, 415ff, 440–42, 448, 1182; W. Arthur, *The Origin of Animal Body Plans*, 158–61, 179; L. Bry, P. G. Falk, T. Midvedt, and J. T. Gordon, "A Model of Host-Microbial Interactions in an Open Mammalian Ecosystem," *Science* 273 (1996): 1380–83; E. Jablonska and M. J. Lamb, *Epigenetic Inheritance and Evolution: the Lamarckian Dimension* (New York: Oxford University Press, 1995); D. L. Nannery, "The Role of the Cytoplasm in Heredity," in *The Chemical Basis of Heredity*, ed. W. D. McElroy and B. Glass (Baltimore: Johns Hopkins University Press, 1957), 136. Also Rosenberg 1985, 101–04.

50. Cf. E. F. Keller, *The Century of the Gene* (Cambridge, MA: Harvard University Press, 2000). Leading candidates for the title "master molecules" are now the *Hox* genes. It is agreed, however, that embryogenesis involves four processes—cell proliferation, cell specialization, cell interaction, and cell movement—and no single command center directs and coordinates any of these processes synchronically or diachronically. In opposing the genocentric view, some have made proteins the primary determinants; cf., e.g., W. Driever and C. Nüsslein-Volhard, "The Bicoid protein Determines Position in the *Drosophila* Embryo in a Concentration-Dependent Manner," *Cell* 54 (1988): 95–104; B. Alberts, "The Cell as a Collection of Protein Machines: Preparing the Next Generation of Molecular Biologists," *Cell* 92 (1998): 291–94. It would be better, though, to give up the picture on which these one-sided views are based.

51. Cf. K. Murphy, *Janeway's Immunobiology*, 8th ed. (New York: Garland Science, 2012), 17, 267–71, 285–90, 515–19, 612f, 663; F. M. Burnet, *The Integrity of the Body: A Discussion of Modern Immonological Ideas* (Cambridge, MA: Harvard University Press, 1963), 14, 37–41, 61, 157f, 172–77, and "Immunological Recognition of the Self," Nobel Lecture 1960 www.nobelprize.org/nobel_prizes/burnet-lecture.pdf.

52. Cf. Lewontin 1983, 63–82, and his 2000, 42–47; F. J. Odling-Smee, K. N. Laland, and M. W. Feldman, *Niche Construction: A Neglected Process in Evolution* (Princeton, NJ: Princeton University Press, 2003); Gilbert and Eppel 2009, 391ff.

53. Buss 1987, 20, 24, 29ff, 52–55, 76–78, 116, 121ff, 170ff. Also Maynard Smith and Szathmáry 1995, 243ff ; Sterelny and Griffiths 1999, 12. That these multicellular systems are real individuals: K. R. Benson, "Biology's 'Phoenix': Historical Perspectives on the Importance of the Organism," *American Zoologist* 29 (1989): 1067–74; B. Santelices, "How Many Kinds of Individual Are There?" *Trends in Ecology and Evolution* 14 (1999): 152–55; J. W. Pepper and M. D. Herron, "Does Biology Need an Organism Concept?" *Biological Review* 83 (2008): 621–27; R. Lewontin 2000, 114. That the origination of eukaryotic protists was similar: L. Sagan, "On the Origin of Mitosing Cells," *Journal of Theoretical Biology* 14 (1967): 255–74; Maynard Smith and Szathmáry 1995, chapter 8. Your body's intracellular mitochondria come originally from your mother, as do its endogenous bacteria.

54. Attempts to establish such a discipline remain marginal. For a sympathetic treatment of them, cf. D. J. Haraway, *Crystals, Fabrics, and Fields: Metaphors That Shape Embryos*, with Foreword by S. F. Gilbert (Berkeley, CA: North Atlantic Books, 2004).

55. Cf. Alberts et al. 2004, 47. Compare Gilbert and Epel 2009, 114, 370, 404, 413f, and D. Haraway, *Crystals, Fabrics, and Fields*, xii, 188–95.

56. Compare Lavoisier, who realized that methodological atomism, which he accepted, did not entail metaphysical atomism, about which he remained agnostic.

57. "What utterly baffles me is why one cannot be a reductionist and a holist at the same time": J. T. Bonner, *The Evolution of Complexity by Means of Natural Selection* (Princeton, NJ: Princeton University Press, 1988), 4.

58. Cf. E. Mayr, E. G. Linsley, and R. Usinger, *Methods and Principles of Systematic Biology* (New York: McGraw-Hill, 1953), 5, J. M. Diamond, "Zoological Classification System of Primitive People," *Science* 151 (1965): 1102–04, S. J. Gould, "A Quahog is a Quahog," *Natural History* 88.7 (1979): 18–26, C. H. Brown, *Language and Living Things: Uniformities in Folk Classification* (New Brunswick, NJ: Rutgers University Press, 1984). Compare Hull 1992, 50f.

59. Cf. R. R. Sokal, "The Continuing Search for Order," *American Naturalist* 126 (1985): 729–49, reprinted in Sober 1994. For critiques, cf. Rosenberg 1985, 183–86; Sober 1993, 163–66; D. Futuyma, *Evolutionary Biology*, 92–94; Sterelny and Griffiths 1999, 186–88. Compare R. Boyd, "Homeostasis, Species, and Higher Taxa," in R. A. Wilson 1999, 141–85.

60. J. Wilson 1999. Some scientific essentialists rely on the genetic approach alone or primarily: e.g., Wilkerson 1995, 117 and context; Ellis 2001, 170; Devitt 2008, 351. Wilson argues persuasively against reliance on any single approach (48–68). I would add that reliance on the genetic approach alone presupposes genocentrism, and genocentrism is no longer assumed in biological science.

61. e.g., P. E. Griffiths, "Squaring the Circle: Natural Kinds with Historical Essences," in R. A. Wilson 1999, 219–26. On the connection between this historical-essence theory and the etiological-evolutionary theory of biotic functions, cf. Rosenberg and Neander 2009, 307–34.

62. Cf. R. G. Harrison, "Linking Evolutionary Pattern and Process: the Relevance of Species Concepts for the Study of Speciation," in D. J. Howard and S. Berlocher 1998, 19–31; J. A. Coyne and H. A. Orr 2004, 46–8, 370–72. Also Wilkerson 1995, 136f.

63. Cf. Coyne and Orr 2004, 18, 33, 37f, 40–3, 70f, and 342f on the relation between hybridization and recombination. Also Boyd 1999, 80.

64. Cf. R. Ueshima and T. Asami, "Single-Gene by Left-Right Reversal," *Nature* 425 (2003): 679; Coyne and Orr 2004, 7, 69, 322; D. E. Soltis, C. J. Visger, and P.S. Soltis, The

Polyploidy Revolution Then . . . and Now," *American Journal of Botany* 101 (2014): 1057–78; and S. Servick, C. J. Visger, M. A. Gitzendanner, "Polpulation Genetic Variation, Geographic Structure, and Multiple Origins of Autopolyploidy in *Galax urceolata*," *American Journal of Botany* 102 (2015): 973–82.

65. Cf. E. Pennisi, "Synthetic Genome Brings New Life to Bacterium," *Science* 328 (2010): 958f.

66. Cf. J. Wilson 1999, 84; Hull 1978, reprinted in Sober 1994, 204–05, with J. L. Mackie, "*De* What *Re* Is *De Re* Necessity?" *Journal of Philosophy* 71 (1974): 554f; also M. Thompson, *Life and Action* (Cambridge, MA: Harvard University Press, 2008), 28f, 33–38, 58–82.

67. The literature on emergence is immense. For a good introduction, cf. Bedau and Humphries 2008. Most of this literature has to do with emergent *properties*. My focus is rather on emergent *continuants*, enduring particulars as distinguished from states, relations, processes, or behaviors. I have already conceded that making this distinction in particular cases can be difficult.

68. The different "levels" may not stack up in a unique way. In string theory, for some specific theory *T* there may be an equivalent theory *T** such that objects regarded as simple in *T* are regarded as composite in *T** and vice versa. So, for instance, in one theory quarks are simple and solitons (monopoles) are quark-composites, whereas in the "dual" of this theory solitons are simple and quarks are soliton-composites.

69. See chapter 6 of this volume.

70. See again P. W. Anderson, "More is Different: Broken Symmetry and the Nature of the Hierarchical Structure of Science," *Science* 177 (1972): 393–96, and R. B. Laughlin and D. Pines, "The Theory of Everything," *Proceedings of the National Academy of Sciences* 97 (2000): 28–31, both reprinted in Bedau and Humphreys 2008.

71. Recall that an effective electron includes its ever-changing cloud of virtual particle-antiparticle pairs, so its compositional and metabolic involvement may in fact be grade II. Recall also this electron has causal powers which belong to it necessarily as long as it exists.

72. See chapters 6 and 7 of this volume.

73. Cf. Russell 1948, 461f; and Quine 1969, 138.

# Epilogue

By submitting ourselves to the gentle guidance of everyday experience and generally accepted opinions we were able to answer the first of our two leading questions: What is it to be a natural kind? A natural kind, I said, is a class whose members are all and only those continuants which exemplify, in virtue of their natures or essences, a single type. Alternatively, a natural kind is a type whose tokens are all and only those continuants which exemplify it in virtue of their natures or essences, and which constitute therefore a single natural class. As types, such kinds may or may not be genuine universals; and as classes, they are specific or generic depending on whether the s-resemblance among its members is exact or inexact.

By requiring that the members of exemplary natural kinds be enduring individuals, or continuants, my account remains strongly rooted in common sense and in the etymology of "natural kind" as a technical term. This has the advantage of distinguishing natural kinds (in the primary sense) not only from mathematical kinds, but also from such natural quantities and relations as mass, charge, force, and distance. Moreover, by linking the concept of a natural kind so closely to the concept of a continuant, my account refuses to free natural kinds (in the primary sense) from the metaphysical and epistemological difficulties attending their members. Continuants are essentially beings-in-becoming; their involvement in the vicissitudes of the material world is grade II, not grade I or grade III; and grade II involvement is difficult to understand. It is also difficult to be sure that one is properly identifying genuine continuants and classifying them in virtue their natures. These difficulties spread, as it were, into natural kinds themselves, making them conceptually and empirically somewhat obscure. This consequence, however, is not a sufficient reason for rejecting my account of what a natural kind is, but rather an additional reason for thinking it is on the right track, or so I argued in chapter 5.

Then, by proceeding under the guidance of scientific theories, we succeeded in answering our second leading question: Are there any natural kinds (in the primary sense of the term). Apparently there are, though few if any of them are the kinds held to be exemplary in atomic physics, chemistry, biology, or in pre-scientific opinion. We therefore accept hypothesis H6: there are, in fact, natural kinds.

H6, along with H4 (there are universals) and H5 (there are continuants), constitute the chief conclusion of my inquiry. All three are philosophical hypotheses. Now, like their scientific counterparts, hypotheses of this sort are all subject to further inquiry and to possible revision or rejection. Reasons for doubting H4 and H5 were noted in chapters 3 and 4, respectively. Pre-scientific reasons for doubting H6 were noted in chapter 5, scientific reasons in chapters 6 and 7. To summarize the latter: First, the concept of emergence requires further elucidation. And even if emergence of some sort has occurred in the course of successive symmetry-breaking events, the reductionist tendency in science makes us wonder whether any *substantial* emergence has yet occurred. Second, it is impossible to track and compare molecular continuants without detecting them, and impossible to detect them without transforming them at the quantum level. It thus remains to be shown that these "free" individuals have not been transformed essentially by our attempts to track them. Third, as we descend from multicellular organisms to field quanta, individuality and countability gradually disappear as laws of nature (in the primary sense) gradually appear. There is evidently some tension between continuant-realism and the nomocentric view of nature that we find above all in mathematical physics. Fourth, as we ascend from quantum fields toward living things, the underlying homogeneity gradually disappears as heterogeneity comes into its own, along with causality; we begin to get powerful particulars different in kind. There is evidently some tension between the nomocentric view of nature and natural-kinds realism.

Finally, natural-kinds realism appears after all to be true because somewhere "between" elementary fermions and living organisms we find well defined classes whose members appear to be enduring individuals. We thus agree with Brian Ellis when he takes some classes of atoms or molecules to be the exemplary natural kinds.[1] Before moving on from this conclusion to other matters, however, let us pause to compare Blackie the squirrel with Fred the free $H_2O$ molecule. The nature of Blackie was evidently inevident in his manifold appearances, we observed, and it seemed reasonable to suppose that our perceptual recognition of him over time could be anchored directly in that nature or essence.[2] We therefore hypothesized that continuant realism is true. Now imagine Fred the $H_2O$ molecule moving about freely. What would it mean to say that Fred's essence is evidently inevident in its manifold appearances? And what reason might we have for supposing that our recognitional awareness of this molecule could be anchored directly in its essence? Such questions remain to be answered. Hence, we cannot yet dismiss once and for all the antirealist's claim that our preferred stereotokens and stereotypes are but idealized particulars and ideal types, all of them constructed rather than discovered.

To defend hypotheses H4–H6, and thereupon develop a detailed theory of natural kinds, we must to go on philosophizing in accordance with theoretical science as well as common sense. Yet let us not ignore our third major resource, namely, those philosophical traditions, ancient as well as modern, which bear on our theme. They too may prove helpful, especially if, as here, our interest is also in renewing natural philosophy in a rather traditional sense of the term. I do make use of this third resource in a forthcoming companion volume entitled *The Aristotelian Tradition of Natural Kinds and Its Demise*. It suffices here to summarize my findings in a statement of three historical facts—facts whereof natural-kinds realists and antirealists, and philosophers of nature generally, would do well to make themselves cognizant.

First, it is a fact that philosophers have not for the most part been natural-kinds realists. Presocratic natural philosophers who came before Parmenides were all monistic mass ontologists (with the possible exception of Heraclitus, who may have been a monistic process ontologist). You might think that Thales, for instance, who took all things to be what he called "water," also took what everything *is* to be one of a kind. But even apart from difficulties surrounding any attempt to regard water as a singleton class, there is no evidence that these pre-Parmenidean philosophers were trying to identify the nature of things in terms of token and type or class and class-member. Although they appear to have been continuant realists of a sort, the very idea of a natural kind was irrelevant to their undertaking.[3]

Presocratic natural philosophers who came after Parmenides were ontological pluralists. Two of them, Leucippus and Democritus, admitted real individuals (atoms), each ungenerable and imperishable, whose relative positions are forever changing. And Empedocles, another post-Parmenidean Presocratic, admitted four distinct masses (air, earth, fire, and water), each ungenerable and imperishable, which mix and unmix cyclically as a result of the operations of Love and Strife. So these three ontological pluralists were also continuant realists of one sort or another. Yet they too appear not to have admitted any natural kinds: there is no evidence that Leucippus or Democritus took atoms to be constitutive of discrete classes in virtue of their natures, and no evidence that Empedocles took earth, water, air, and fire to be four tokens of the type *enduring mass*. Empedocles may have been the first Presocratic philosopher to admit what Homer called "tribes" or "races" of things; but he seems not to have regarded these things (mixtures) as genuine continuants; and even if he did, there is no evidence that he took the swarms they form to be disjoint resemblance classes in virtue of their diverse natures.[4] The concept of a natural kind may be hovering in the wings, but it does not yet have any theoretical role to play.

Plato followed Parmenides in distinguishing strictly between being and becoming.[5] And he followed Socrates in turning philosophy away from any direct investigation of nature, and to a close investigation of things human. He also followed Socrates in positing not one but several beings, called "forms" or "ideas," and trying to say what each of them is. These ungenerable and imperishable forms are universals, as Aristotle noted; and insofar as transient things "partake" of these universals, we note, they instantiate or manifest them. It is in Plato's writings that we find, for the first time, a distinction between real types and their tokens.[6] Yet Platonic forms (*eidê*) are not natural-kind types, since the things partaking of them are not *beings*-in-becoming; and although the different groups (*genê*) into which these quasi-continuants fall may seem to us to be natural-kind classes, they are not well defined apart from the classificatory activities (divisions and collections) of someone's mind. The range of natural *genê* thus resembles a stream of verbal utterance: though articulate in some sense, it is not composed of definite syllables apart from our alphabetization or classification of it.[7] Plato's *Timaeus*, the only dialogue traditionally said to be about nature, only seems to disconfirm this claim. For while it is true that the elementary bodies making up the cosmos form disjoint resemblance classes in virtue of their different geometrical structures, these structures are the result of the Demiurge's geometrization of the pre-cosmic powers existing in Place. Such powers are bodies in their natural state, we could say, before the Demiurge civilized them. And in their natural state, Timaeus argues, they constitute an indeterminate multitude. Indeed, they are not even bodies.[8]

Plato's challenge is to show that *beings*-in-becoming exist, and that some, if not all, of these entities fall into discrete resemblance classes in virtue of their natures. Aristotle took up this challenge. He first modified the Socratic turn by taking what we perceive, not just what we think or say, to be heuristically valuable.[9] He then proceeded to investigate the natural world, always taking his bearings by how things natural appear to us, or by how we ordinarily apprehend them in opinion or speech. Thus did he find that there are naturally constituted continuants, that each of them is an enduring individual in virtue of its intrinsic form (*eidos*), that some of them beget still other individuals having very similar intrinsic forms, and that the kinship groups to which these homo-eidetic individuals belong are in truth disjoint species (*eidê*). Whether Aristotle thought that these typological species$_t$ may also be regarded as genuine universals is a much-disputed question, yet clearly he admitted s-resemblance classes whose members are individual continuants, and such classes are natural kinds in the primary sense of the term.[10]

Aristotle was the first natural-kinds realist in ancient Greek philosophy, and for the next few centuries he was just about the only one. Most natural philosophers in late antiquity endorsed either Stoic physics (which was rather Heraclitean) or Epicurean physics (which was rather

Democritean), and therefore had little need for the concept of a natural kind, save as a pragmatic classificatory device. Only in the so-called middle ages did the Aristotelian view predominate. Then most philosophers were Aristotelians of one sort or another; and natural-kinds realism remained part and parcel of the theocentric cosmology they endorsed.[11] Yet eventually things began to unravel. For instance, William of Ockham undertook to purify both theology and physics by removing from them all Neoplatonic elements. The result was antirealism with respect to universals, including natural-kind types, and anti-essentialism with respect to singular things, including putative continuants. The latter consequence, in particular, threatened to undermine Aristotelian natural-kinds realism altogether; for if no distinction at all can be made between a continuant ($x$) and its essence or nature (n$x$), there is no longer any basis in reality for saying of two continuants $x$ and $y$ that although they differ materially they nonetheless resemble each another essentially if indeed they are homo-eidetic.[12]

Early modern natural philosophers followed Ockham in denying the reality of natural-kind types. They also denied the reality of "substantial forms." Consequently, for them the Aristotelian distinction between heaps and naturally constituted wholes has some basis in reality only if some materially particles are genuinely atomic. Though a few did endorse metaphysical atomism, most did not, preferring instead to suppose that the natural world is at bottom just matter in motion—a view remarkably like that of the Presocratic natural philosophers before Parmenides.[13] Some modern philosophers balked. Leibniz, for one, insisted that the dynamism of the world can be fully explained only if it contains powerful particulars that differ essentially from heaps. But his version of continuant realism never caught on, and what he calls "species" are better understood as phase sortals (like *tadpole* or *caterpillar*) than as natural kinds in the primary sense of the term.[14] Locke, though not primarily a natural philosopher, did allow for corpuscular essences that are not merely nominal. But these real essences are beyond our cognitive reach, he said, and in any case too disunited and too transitory to be the essences of genuine continuants. It is always in terms of nominal essences that we sort material things into discrete kinds, and arguably in terms of such essences that we even classify some material things as individual continuants. The worth of such classifications, he claimed, is at bottom practical rather than theoretical.[15] Kant agreed. We find it useful in scientific research, as in daily life, to provide ourselves with ever more refined typologies of things; but to become truly scientific, he argued, these typological descriptions of nature must give way to an explanatory history of nature in which any remaining sortal terms are but mnemonic devices or the like.[16] Kant also maintained that, in order to understand living things, we *must* regard them *teleo*-mechanically as *organized* beings. Yet this a priori necessity, he insisted, is methodological ("regulative") rather than constitutive. Thus his account of living things only seems Aristotelian.[17] Whether this regulative principle is even necessary *for us* was put into doubt by Charles Darwin, who showed not only how one can explain the origin of species (species$_g$) in a wholly naturalistic (mechanical) fashion, but also how all of biological theory can be brought within the evolutionary framework he has outlined.[18]

Natural philosophers and scientists today tend to be natural-kinds antirealists. Yet some are not. Those called "scientific essentialists" have tried to bring molecular chemistry and particle physics, in particular, within the new tradition of natural kinds instituted by Mill and renewed by Kripke. Their views, however, are not yet widely shared. Natural-kinds realism remains a minority position in natural philosophy.

From this fact, of course, it does not follow that natural-kinds realism is incorrect, but it should not therefore be ignored by natural-kinds realists. Similarly, from the fact that natural-kinds realism is in greater agreement with common sense than any of its major alternatives it does not follow that these alternatives are all incorrect; but it should not therefore be ignored by Heracliteans, Epicureans, Cartesians, and Newtonians, or by general relativity and quantum field theorists. Indeed, the very tendency of natural philosophers to think that the world as it is little resembles the world as we commonly understand it—a tendency constrained to some extent by Aristotle's neo-Socratic procedure, by the scientist's experimental method, or by Kripke's way of appealing to everyday "intuitions"—is something that should interest all of us. Are there any a priori constraints on how far natural philosophy can go before its account of nature becomes unacceptable? There are a few. One is that the account be intelligible, that it not be formally contradictory or conceptually incoherent. If a philosopher denies that motion is possible (as Zeno did), has she in effect removed nature itself? So it appears. What if she admits the metaphysical reality of nature while denying its protometaphysical reality? Then too she has gone too far, or so I argued in chapter 2. But what if she claims that mathematical physics is reducible in principle to pure mathematics? This Cartesian option is harder to dismiss once and for all: though rejected by Leibniz and Newton alike (and by me in chapters 3–5), it has reemerged with the advent of general relativity theory, superstring theory, and ontic structural realism. Or what if she maintains that quantum reality so differs from our everyday world that it will forever baffle our attempts to understand it? This claim flies in the face of our desire to know, but cannot be dismissed out of hand. Or, finally, what if she agrees with Husserl that modern mathematical physics constitutes a pyrrhic victory for human reason—a victory inasmuch as the physical world now appears subject to mathematizable laws, pyrrhic inasmuch as human reason can find no place for itself or its activities in a world so conceived?[19] But this has been a perennial problem for natural philosophy. Descartes saw no way of reducing himself, as a thinker or as a human being, to something he regarded as nothing more than a delimited portion of space. And Thales, I imagine, did not succeed in reducing himself, or his thoughts, to the stuff he called "water." Does the presence of this problem render an account of nature unacceptable? No, for it it may only seem insoluble, or it may not be a problem that *natural* philosophy has to solve or diffuse. Husserl did not conclude that "Galilean science" is theoretically unacceptable, nor did he imply that our returning to the *Lebenswelt* as a field of phenomenological investigation would lead us to conclude that natural-kinds realism is after all correct.

Second, it is a fact that, for us, the greatest "paradigm shift" in natural philosophy remains the one in which Aristotelian physics was replaced by Galilean science. According to Aristotle, natural substances are to be explained by reference to their intrinsic forms (*eidê*). And several such entities are one in kind (*eidos*) if and only if they resemble one another very closely in virtue of their forms. There is in his physics no appeal to law (*nomos*). According to Galileo, natural phenomena are to be explained by reference to the intrinsic laws governing them. There is no appeal to substantial forms. If in his physics there should be any mention of essentially different species of *things*, these are but heuristic devices or concessions to everyday understanding. So whereas Aristotelian natural science is eidocentric and admits natural kinds, Galilean physics is nomocentric and admits no natural kinds.[20]

Aristotelian species are explicable by real definitions. Let "Man is a rational animal" be such a definition. The term "man" designates the species whose members or instances (specimens) are all and only human beings, while the expression "rational animal" identifies the genus to which these continuants belong and the specific difference that sets them apart from all other continuants belonging to the same genus. "Man is a rational animal" tells us what each and every human being is essentially; it declares what is invariant in human beings as beings, however various these enduring individuals may be at any given time or over any given period of time. Laws of nature, on the other hand, are expressible as equations or proportions. Let "$S \propto t^2$" be such a law for bodies in free-fall near the surface of the earth. The terms "$S$" and "$t$" designate distances traversed and time elapsed—magnitudes which appear to be quite different in kind—and the proportionality sign tells us that if a body falls one unit of distance in one unit of time, it will fall four units of distance in two units of time, nine units of distance in three units of time, and so on indefinitely. We see that something is invariant in all these different cases: it is the same duplicate ratio, as they used to say, or the same functional relationship, the same equation, as we now say. "$S \propto t^2$" declares an invariant relation between distances and times, however greatly these magnitudes may vary.

Though Aristotelian and Galilean natural philosophers all try to discover what in invariant in the welter natural variations, the difference between Aristotelian natural kinds and the Galilean laws of nature is immense. Regarded logically, the former are like terms or concepts, the latter like statements or propositions. Regarded ontologically, the former are sortal types or classes whose instances or members are s-resembling substances, whereas the latter are relations or relational types, not necessarily causal, whose terms are physical quantities. The difference between an individual continuant ("primary substance") and a determinate quantity is categorial. So too is the difference between a species ("secondary substance") and a nomic relation. Aristotelian physics features heterogeneous substances and their kinds, whose quantitative and relational attributes it takes to be nothing more than dependent modes. Galilean physics does not. To the contrary, it features causal and noncausal relationships among physical quantities of different sorts (masses, lengths, times, etc.), whose bearers, *if any*, are reducible either to homogeneous matter or to atomic particles that do not differ in kind. The *essence* of free-fall motion, Galileo said, is given by the formula $S \propto t^2$. Absent complicating circumstances, the falling bodies may be stones, glass beads, raindrops, clams, or human beings; the evident differences in kind among these mobile particulars make no difference to the motion itself.[21] The world according to Galilean natural philosophers is essentially mathematical, or mathematizable, whereas according to Aristotelian natural philosophers, it is only locally and incidentally so. The new mathematical physics is therefore to be distinguished sharply from such pre-modern sciences as optics and astronomy, in which Euclidean geometry was shown to be applicable to restricted domains of natural phenomena that were presumed to be essentially non-mathematical, namely, those bodies said to be transparent or reflective, and those heavenly bodies whose every movement was thought to be uniformly circular.

Aristotelians also admit cases of substantial change in nature—the coming into being or perishing of a human being, for example. Galileans do not acknowledge any substantial emergence or submergence other than divine creation or annihilation. Like the post-Parmenidean Presocratics long ago, they reduce all natural change to locomotion. Aristotelian physics also posits a hierarchy of entities, ranging from elemental masses (earth, water, air, and fire, in that order) to living things, from plants to animals, from sea cucumbers and barnacles to horses and human beings, and from terrestrial things to things celestial; and it explains many, though not all, of these things in terms of final causes. Galilean physics admits no final causality and posits no such cosmic hierarchy; to the contrary, it further homogenizes nature by putting all bodies on a par. Even the distinction between naturally constituted things and artifacts is erased.[22] For Galilean philosophers, the fundamental distinction is not that between what is by nature and what is by art, but between natural motions, which are thoroughly determined, and voluntary actions, which presumably are not. The old distinction between nature and art (or convention) has given way to a new, very different distinction between nature and freedom (or history).

The difference between Aristotelian and Galilean physicists is not simply theoretical. They also have different starting points, different ways of proceeding, and different goals. Both begin with things that are familiar or first for us; but whereas Aristotle takes these to be exemplified by such living things as horses and humans, Galileo takes them to be exemplified by such mechanical devices as pendulums and pulleys. His natural philosophy is mechano-centric. It focuses initially on things that are inanimate and apparently rather simple, whose characteristic motions seem to be reversible (you can "run the movie backwards" without violating the functional invariance in question). Aristotle's natural philosophy is rather zoö-centric. It focuses initially on things that are animate, apparently very complex, whose characteristic motions seem to be irreversible (embryogenesis and aging, for example). One could hardly choose more disparate phenomena to represent what is first for us.

Whereas Aristotle's physics is first of all zoology, Galileo's is first of all terrestrial mechanics. Both ground their explanations empirically, and yet the ways in which they gather their empirical evidence are quite different. For Aristotle, it was primarily by means of hands-off observations of living things in their natural habitats, whereas for Galileo, it is primarily by means of experiments in highly artificial situations.[23] For Aristotle, precise quantitative measurement was of little or no importance, whereas for Galileo it is crucial. And, having thus established explanatory frameworks for their preferred phenomena, they both try to generalize their accounts to include everything natural. But for Aristotle this meant proceeding downward, as it were, to take into account inanimate bodies or masses, and upward to take into account celestial phenomena, whereas for Galileo it means extending his mechanical explanation to celestial phenomena and to living things. And whereas Aristotle found that his four-fold explanatory scheme could not be extended to inanimate things without modification,[24] Galileo supposes that his explanatory scheme can be extended to living organisms without any modification at all; for again, all bodies are on a par; the only difference is one of complexity, and anything naturally complex can be analyzed into epistemically simple elements from which it can then be reconstructed in thought and in fact.

Finally, while the goal of Aristotelian and Galilean natural philosophers is an adequate theory of nature, they mean very different things by "theory." Apart from some animal dissections, Aristotle did not intentionally intervene in what he was studying. Instead, he watched. Theory (*theôria*) for him was spectatorial rather than practical or productive; it was not the sort of thing one might construct. Galileans, on the other hand, do intervene methodically in nature—or as Bacon so colorfully put it, they torture her, in order that she might reveal her secrets. For them, moreover, a theory *is* something constructed. In fact it is a work in progress, one that builds on collateral developments in technology, mathematics, and civil society. Thus have these new scientists elided the old distinction between theory and practice or theory and production. To Aristotelians they appear fundamentally confused. To Galileans, however, the old spectatorial conception of theory appears much too narrow. It makes theoretical science ineffectual by depriving it of the power that comes with knowing how to produce what one has explained.[25]

So different are Galilean physics and Aristotelian natural science as Kuhnian "paradigms" or "disciplinary matrices," and so persuasive has the Galilean paradigm been for us, that any return to Aristotle's eidocentric view of nature seems out the question. Galilean physics, moreover, strongly suggests that the concept of a natural kind has no theoretical use; that putative natural kinds, and even enduring individuals, will disappear in the final theory of everything. What, then, are natural-kinds realists to do? It would be foolish for natural philosophers today to ignore contemporary scientific theory, as it may turn out to be our most valuable heuristic resource. No less foolish would it be simply to assert, in the teeth of evidence to the contrary, that contemporary scientific theory entails the existence of natural kinds in the primary sense of the term. A much better response, I believe, is to begin as the early Galileans did, by taking apart the total package that is the prevailing disciplinary matrix, and examining what appear to be its weak points. For just as the Aristotelian acceptance of definable species$_t$ does not require a spectatorial conception of theory, so too the Galilean acceptance of mathematizable nomic relations does not require an instrumentalist conception of theory. And just as Aristotle's eidocentric conception of nature does not, strictly speaking, exclude the possibility of a mathematical mechanics that is indifferent to substantial differences in kind or the possibility of a nonteleological and nonhierarchical cosmology, so too Galileo's nomocentric conception of nature does not strictly preclude the existence of such differences in kind, or of the "thick" particulars which may be members or instances of such kinds. There are natural kinds in the primary sense, I have argued, only if there are genuine continuants. Galilean science strongly suggests, but does not entail, that there are no cases of *substantial* emergence, no *beings*-in-becoming, that every enduring *thing* must go. This is a weak spot. Those natural philosophers who would be natural-kinds realists should focus first of all on it—just as early Galileans focused not on Aristotle's biology, which still looked strong, but on his mechanics and his cosmology, which seemed unduly qualitative and speculative.

As we have seen, however, those who wish to find a place for continuant-realism in contemporary scientific theory have a hard row to hoe. On the one hand, some living organisms appear to be enduring individuals; and it was by focusing on them in chapter 4 that I was able to propose continuant-realism as a hypothesis. But then I was proceeding under the gentle guidance of common sense. Had it been under the guidance of biological theory instead, as in chapter 7, the ontic status of living organisms, not very clear to begin with, would have been still more obscure. I might then have proposed continuant-antirealism instead, or else refrained from accepting either hypothesis.[26] The problem here is that while ordinarily, such continuant-like particulars do appear to have natures in virtue of which each remains the same as long as it exists, biological theory recognizes no such natures or essences. Consequently, though it admits some living things as stereotokens, these particulars appear on reflection to be mere *pragmata,* and the corresponding stereotypes it proposes—whether morphological, genetic, or ecological—seem merely pragmatic; none appears on reflection to be a real natural kind ($species_t$).

In physics, on the other hand, we do find real heterogeneity. This is because physics today, while still fundamentally Galilean, has rejected the longstanding supposition that all physical quantities are continuous. Electric charge, for example, is now thought to exist only in discrete units (quanta): every electron has exactly one unit of negative electric charge, every proton exactly one unit of positive charge; hence every atom or ion has exactly $n$ units of negative charge and $m$ units of positive charge, etc.[27] Some natural-kinds realists have seized upon this fact, though not always with sufficient circumspection. For again, when applied to two electrons $a$ and $b$, Coulomb's Law ($F_e \propto e_1 e_2/r^2$) concerns not $a$ and $b$ directly, but rather the quantities of charge $e_1$ and $e_2$ that they carry. These charges are quantitatively identical (equal). So too are masses and spins of the particles that carry them. Moreover, these three properties are thought to be essential to any electron: essentialism has disappeared from biology and reappeared in particle physics, a development that would have astonished earlier Galileans. But again we need to be careful. Electrons are not individuals in the usual sense, and arguably not *things* at all but rather normal modes of the quantum field to which they belong. To show that these "particles" constitute a natural kind in the primary sense, one must show that the genesis of an electron is a case of *substantial* emergence within the quantum field on which it depends, or that a real (not merely virtual) electron is a "thick" particular that remains the same over time, and not merely a "thin" particular that can exist only for a instant. And this, we found, is not easy to show.[28]

To secure a place for natural-kinds realism within scientific theory, the most promising course is right up the middle, into theoretical chemistry. Consider once more the class of $H_2$ molecules. These chemical steretokens are indeed continuant-like things. Are they genuine continuants? It would be difficult to show that they are not. And if they are, do they not then constitute a natural kind in the primary sense of the term? It would be difficult to show that they do not. Yet even here there are problems. It suffices here to recall some things noted in chapter 6. With regard to the natures of a molecule's chemical constituents, we confronted the potentiality problem, the isotope problem, and the decay problem, and found that they were difficult to resolve if one denies that the chemical "natures" of these constituents exist only in relation to chemical practice, that the corresponding chemical kinds are pragmatic rather than natural.[29] Also, with regard to the natures of the molecules themselves, we should take our bearings by the relevant scientific stereotypes, since these presumably describe what the corresponding stereotokens are essentially. For $H_2$ molecules, the most accurate and insightful stereotype is a properly specified wavefunction, or so I argued in chapter 6.[30] You may disagree. You may prefer something more traditional, along the lines of "Water is hydrogen oxide or $H_2O$." Be that as it may, there is still a problem, for one must show that your preferred theoretical stereotype, unlike those we employ in everyday life, really does declare what the molecular continuants so defined are essentially.

Third, it is a fact that natural philosophy has always been an iffy endeavor. This becomes evident on a personal level when you reflect on your own philosophical investigations of nature. Suppose, for example, that you are trying to discover whether time is irreducibly asymmetrical, as laymen and scientists alike tend to believe. You then realize that you do not know the answer to your question, and that no one can give you the knowledge you seek; you must find out for yourself. Nor can others tell you how to look for the answer you seek, since any methodological opinions they might have would be just that, opinions, all in need of examination by you. Philosophizing is necessarily first-person, even when one is cooperating with others. While philosophizing, moreover, you are always between two stools, no longer resting content with some given answer to your question but not yet knowing what the correct answer is. And insofar as you cannot deduce that answer from what you already know, your inquiry is epistemically nonfoundational; you are proceeding without much of a safety net. Philosophizing is necessarily a matter of trial and error. Like Eros according to Plato's Diotima, your inquiry is an offspring of poverty and resource, and like Eros it is liable to collapse and renewal.[31] You may at any moment feel a need to digress, to examine something that you had taken for granted, to consider some unexpected difficulty, or to begin again from the beginning. One of my aims in this book has been to display and make explicit these characteristics of natural philosophy as an activity.

The instability of natural philosophy also becomes evident when one reflects on its history from Thales to the present. I count four major crises, whose prime movers were Parmenides, Socrates, medieval Schoolmen, and Kant. I shall now state the challenge each posed, and briefly describe how natural philosophers responded to it.

According to Parmenides, the very idea of nature is incoherent inasmuch as it rests on a refusal to distinguish strictly between being and nonbeing. Natural philosophers unwittingly exemplify the way "on which of mortals who know nothing wander double-headed, since helplessness in their breasts guides their vagrant mind they are carried deaf and blind alike, dazed indecisive tribes for whom being and nonbeing are held to be the same and not the same."[32] Natural philosophers posit two "forms," Light and Night, in terms of which they try to understand everything that comes into being and perishes. (Heraclitus called them "the logos" and "everliving fire"; Philolaus, "limit" and "unlimited"; Aristotle, "form" and "matter"; Newton, "deterministic force laws" and "inertial matter"; quantum theorists, "mathematical structure" and "indeterminate field disturbances.") The problem for these philosophers is then two-fold: to elucidate Night without explaining it away (as Descartes tried to do), and to understand how Light and Night can be both two and one (as Aristotle tried to do) or how there can be beings-in-becoming, things who involvement in the ever-changing world is grade II and not grade I or grade III (as I have tried to do). This problem is irresolvable, Parmenides implies, not because nature likes to hide (as Heraclitus said) but because the whole enterprise is fundamentally misconceived. There is no such pair as Light and Night. What comes to be and perishes only *seems* to be. Philosophers should therefore turn away from the natural order, which rests ultimately on the arrangement of their words, and endeavor to place themselves on the nonnaturalistic way of truth, which is about being and nothing else.[33]

How Presocratic natural philosophers critiqued the Parmenidean critique is unclear. What is clear, however, is that this challenge persuaded them to replace monism with pluralism, reject all self-movement, and deny that there is any sort of change besides combination and separation. They also admitted irreducible qualitative differences among everlasting individuals or masses, thereby raising the prospect, at least, of a physics in which there are essential differences in kind.

The second great crisis in natural philosophy was instigated by Socrates. He was, according to Plato's *Phaedo*, a natural philosopher who, having found himself falling deeper and deeper into perplexity, turned away from sensory experience and becoming to arguments based on hypotheses about what things *are*.[34] Yet the Socratic Turn was more radical than this description of his "second sailing" suggests. For Socrates maintained that the proper study of man is man; that philosophers should concentrate not on the natural world (into which they are wont to read themselves) but on the human soul, the standards by which we judge one another and the different societies in we live, and the ultimate end we all strive to attain. Socrates was thus the first to admit universals, which he tried to define in tense conversation with other interested parties. He may also be said to have founded moral or political philosophy. Notice, however, that political philosophy for him was not a branch or division of philosophy generally. He suggested rather that political philosophy *is* philosophy, or at any rate *first* philosophy, whereas natural philosophy is not. Hence any direct investigation of nature, though perhaps not wrongheaded in the way Parmenides alleged, is nonetheless a distraction for philosophers as such. Plato did not disagree. Even Timaeus, his exemplary natural philosopher, says that cosmology is engaged in for the sake of relaxation from the rigors of ontology, and that it is about becoming rather than being, and about becoming there can be no knowledge but only opinion with sense perception.[35] Plato, moreover, has set this exemplary cosmology in a context that is overtly political: when Socrates has finished summarizing his city in speech and asked that it be put in motion, Critias orders Timaeus to describe the cosmos in which such a city has presumably come into being. So even if natural science were possible, the ostensible purpose of Timaean cosmology is not to outline such a science but to mediate political theory (the summary account drawn from Plato's *Republic*) and political history (the impending account of Atlantis in Plato's *Critias*).[36] What looks like an exhibition of natural philosophy has been made part and parcel of an exercise in political philosophy broadly conceived.

To this challenge, natural philosophers responded in various ways. The Epicureans mostly ignored it, preferring instead to refine and elaborate Democritean atomism.[37] The Stoics agreed with Socrates that the proper study of man is man, and with Aristotle that physics and ethics are distinct parts of philosophy; but unlike Socrates or Aristotle, they chose to elaborate their physics within a Heraclitean framework.[38] It is Aristotle's physics that owes the most to the Socratic-Platonic Turn from which it most explicitly dissents. Methodologically, it retains their insistence on beginning always with what is first for us, while rejecting their denigration of perceptible phenomena and becoming. Metaphysically, it retains their attention to definable forms (*eidê*), while rejecting the claim that such forms must transcend all things changeable. The transition from Presocratic to Aristotelian physics is arguably the first of two "paradigm shifts" in the history of natural philosophy.

The third major crisis was initiated by those medieval Schoolmen who subordinated natural philosophy primarily to the authority of Church doctrine, and secondarily to the authority of Aristotelian physics. So, for example, since creationism is true according to Church doctrine (and common opinion), no philosopher should think otherwise; and since the world is hierarchically ordered according to Aristotle (and common opinion), no natural philosopher should think otherwise. That philosophy cannot be the handmaiden of any authority outside itself seems not to have occurred to these Schoolmen. For them, the study of nature consisted mainly in thinking through and commenting on the relevant Aristotelian treatises, together with extant commentaries thereon, sometimes with a view to showing how Aristotelian physics can be brought into agreement with Church doctrine, sometimes with a view to showing how they disagree.[39] They saw no need to begin again from the beginning, or to use Aristotelian physics and Church doctrine as heuristic resources rather than final authorities, in a sustained effort to find out for themselves how things natural are to be understood.[40]

Seventeenth-century natural philosophers responded to this challenge by tacitly freeing philosophy from authoritative pronouncements of the Church, which they continued to obey insofar as they were Christians, and by openly rejecting Aristotelian physics altogether. I have already described the new science they established in its place. It suffices now to note that, in some respects, what we today regard as physics is a return to Presocratic physics. Bacon himself noted as much, and more recent historians have described in detail the Lucretian (Democritean) and Archimedean (Pythagorean) character of the new mathematical mechanics.[41] The second "paradigm shift" in the history of natural philosophy is therefore, in a way, a reversal of the first. But only in a way; for with respect to its methods and goals, its insistence that there be laws governing all natural phenomena, and the still-unresolved difficulty it confronts when trying to find a place for man in a world governed by such laws, the new mathematical mechanics could hardly differ more from those pre-Christian and non-Socratic precursors.

The fourth major crisis in natural philosophy was instigated by Kant in the *Critique of Pure Reason*. His challenge was two-fold. On the one hand, he argued that only if we regard the natural world as empirically real but transcendentally ideal can we secure the foundations of Newtonian physics against Hume's skeptical attack, resolve some otherwise some intractable antinomies in metaphysics, and find room for moral freedom in a deterministic world.[42] But if nature is transcendentally ideal, if its very reality depends on the *a priori* structure of the human mind, then all earlier physicists, and many since, have been fundamentally mistaken. On the other hand, Kant reinforced the tendency of Galilean science and philosophy to go their separate ways, by maintaining that the investigation of nature (empirical reality) is the job of scientists alone. Philosophers may help them by clarifying what it is to be a science, by defining the concepts proper to science, and by anticipating what "total science" will look like; but that is all.[43] There is no longer any place for natural *philosophy*. It has been replaced by natural science and the philosophy of science.

Although Kant's transcendental idealism is no longer accepted, the two-fold challenge he posed lives on, having metamorphosed into the two mutually independent doctrines I have called "protometaphysical antirealism" and "philosophical naturalism." How is one to renew natural philosophy in the face of this double threat? Some try to show that neither poses an insurmountable obstacle, as I did in chapter 2. Others choose to ignore both: they are unabashed protometaphysical realists who deny that philosophy is, or ought to be, a handmaiden of science. It does not follow that they dismiss scientific theory. Some, in fact, use it as their principal heuristic resource, while others prefer to use it in combination with common sense, as I have, and still others prefer to rely on common sense alone. The results of their post-Kantian metaphysical investigations of nature are remarkable for their variety. To mention only those most relevant to my present inquiry: there are now realists and antirealists with respect to natural kinds, continuants (endurants), four-dimensional occurrents (perdurants), and laws of nature; those who maintain that properties are universals and those who do not; those who maintain that some physical properties are essentially dispositional and those who deny it; those who deny that elementary particles are just normal modes of the quantum fields to which they belong and those who affirm it; those who affirm that animal kinds are really just $species_g$ and those who deny it. Which, if any, of these doctrines will prevail in the natural philosophy of the future? Who knows? And will any of these philosophical developments catalyze another paradigm shift in science? Again, who knows? For the time being, at least, it seems more fruitful to think of ways in which natural philosophers and natural scientists might utilize each other. I have already suggested that philosophers who believe that substantial emergence occurs, and scientists who believe that a theory of everything is not a theory of every thing, would do well to undertake a close collaborative study of symmetry-breaking events.

You will have noticed that, although I count myself among those who are renewing natural philosophy today, this book is short on doctrine. Certainly it does not present a full-fledged theory of natural kinds. Yet it does succeed in clarifying the concept of a natural kind, I believe, and revealing its distinctive role in our thought. It also offers reasons, drawn from scientific theory rather than common opinion, for believing that natural kinds exist. And by presenting all this in a somewhat roundabout way, it has drawn a map, as it were, of the very rugged conceptual landscape on which all natural philosophers today must proceed whether they like it or not. This map is necessarily incomplete, inasmuch as my focus has been on natural kinds and not, for example, the nature of space-time, causality, or laws of nature. It may also be in need of revision even with respect to natural kinds and the things which constitute them. It is then but a draft of a much-needed map, offered here so that others may use it as a guide in drawing one that is more accurate and less incomplete. I invite you to try.

## NOTES

1. Ellis 2001, 3, 161–65, and 2010, 58–60, 106, 112–14. Thus, whereas Aristotle's natural-kinds realism was bio-centric, Ellis's is thus chemo-centric. See also Psillos 2002, 288 and context; Slater 2005, 31ff.

2. See chapter 4 of this volume.

3. H. Diels and W. Kranz, *Die Fragmente der Vorsokratiker* (Berlin: Weidmann, 1951–1952), 11A12, 12A9, and Heraclitus B30, 50. All my references to Presocratic fragments are according to the enumeration in Diels-Kranz.

4. Leucippus B2, Democritus A6, 8, 11, 13–14, 37, Empedocles B6–7,9.2, 11–14, 17.4–28, 21.9, 14, 22.4–5, 26.3, 6, 12, 30.1, 35.7, 11, 62.4, 98.1–2, 109, 124, 129.5

5. In Plato's *Sophist* the Eleatic Stranger rejects the famous "Platonic" separation (*khôrismos*) of being from becoming; and he must do so, it seems, if he is ever to capture the sophist in a definition. To discuss this notable exception here would take us too far afield.

6. Plato, e.g., *Euthyphro* 6d9–e6, *Meno* 72c1–d1. Note that when asking and pursuing a question of the form "What is X?" Plato's Socrates often identifies X as an *eidos*, never as a *genos*. The word *genos* is cognate with words that signify generation, whereas the word *eidos* is cognate with words that signify knowing or seeing. They belong to two very different semantic fields.

7. Plato, *Cratylus* 393d1–e9, *Philebus* 18a7–d2. The *Philebus* is the only Platonic dialogue in which Socrates mentions natural kinds as types; but he calls them "henads" (units) rather than "forms," and their ontic status is left undetermined.

8. Plato, *Timaeus* 50a3–51b2 and 52d-6161c, especially 53a2–e7, 54b3–d7, 55a2–56d1, 57c7–d6.

9. Aristotle, *Nicomachean Ethics* 1145b2–20, 1146b8ff, *Topics* 100a18–b23, 104a8–11, 108b13, 119a38, *Prior Analytics* 24b12, *Art of Rhetoric* 135b27–32, *de Caelo* 270b6–20, 279b6–7, 287b31–288a2, 297b23–298a20, 308a4–7.

10. Aristotle, *Physics* 193a4–b13, *de Anima* 412a11f, *Metaphysics*, 984a30–b1, 992b18–20, VII.3–IX.8, *Categories* 2a14ff, 3a7–25, b10–21, *Generation of Animals* 715a23, b3f, 734a28–b22, 767b14.

11. E.g., Thomas Aquinas, *Summa Theologica* I.3.8ad2, 4.3, 14.11, 15.1–2, 29.2ad3, 34.3ad4, 44.3, 66.1, 70.3, 75.4, 76.3ad2, 84.4–5, 113.2, II.93, *Summa contra Gentiles* 2.12, 3.24.

12. Ockham, e.g., *Quodlibetal Questions* 2.7, 10–11, 4.19, 21, 5.11, 15, 5.23, 6, 6.15; *Summa Logica* 1.1, 4, 13–19, 33, 42; *Ordinatio* 1.2, *Epositio Physicorum* 1.1, 16, 2.2, 9. Notice that he does not reaffirm the Aristotelian distinction between heaps and naturally constituted wholes: when a piece of wood is cut in two, he says, one substance becomes two substances.

13. E.g., Descartes, *The World* 5, 9, *Principles of Philosophy* 3.44, 3.46, 52ff, 4.4–6, 35, 198, 203; Bacon, *New Organon* 1.15–16, 43, 66, 2.17; Boyle, *Origin of Forms and Qualities* Preface, 4, 7, *The Sceptical Chymist* (New York: Dutton, 1937), 11–27, 63, 94–112, 132–37, 152, 176, 179, 186f, 223–28; Newton, *Principia*, first ed., Preface.

14. Leibniz, *New System of Nature, On Nature Itself, Principles of Nature and Grace* 3–4, 12, *Monadology* 62–63, 67, 70; "The Human Body" and "On Botanical Method," both in J. Smith, *Divine Machines* (Princeton, NJ: Princeton University Press, 2011), 291, 305.

15. Locke, *Essay Concerning Human Understanding*, e.g., 1.1.6, 2.2.2, 2.8.9, 2.8.13, 2.8.25–26, 2.9.8, 2.10.20, 2.12.3–6, 2.20.6, 2.23.3, 2.23.14, 2.23.23–26, 2.27.1–11, 2.31.1–16, 2.32.7, 3.1.1–8, 3.3.6, 3.3.14–19, 3.5–6, 3.11.7, 3.11.19–22, 4.6.11, 4.11.8.

16. Kant, *Critique of Pure Reason* A228f/B281, A651/B679, A654/B682, A657/B685, A663–64/B690–92, A666f/B694f, *Metaphysical Foundations of Natural Science* 3, *Critique of Judgment* Introduction, 79, 82 *Anthropology from a Pragmatic Viewpoint* 34.

17. Kant, *Critique of Judgment*, 61, 64–68, 70–71, 75, 77–81. See also "On the Use of Teleological Principles in Philosophy" and *Opus Postumum* 21:183–84, 291, 22.355–56, 383, 499, both in *Kants gesammelte schriften*, hrsg. Von der Königlich preußischen Akademie der Wissenschaften (Berlin: G. Reimer, 1910– ), vols. 8 and 21–22.

18. Charles Darwin, *On the Origin of Species*, A Facsimile of the First Edition (Cambridge, MA: Harvard University Press, 1964), 13, 44f, 52, 127–29, 282, 329–36, 343f, 411, 413, 420, 448f, 456–58, 478, 484–86.

19. Husserl, *The Crisis of European Sciences and Transcendental Phenomenology*, tr. D. Carr (Ann Arbor: Northwestern University Press, 1970), §§ 9i, 34d et passim. Compare Nietzsche, *Beyond Good and Evil*, §§ 9, 22.

20. For a similar "paradigm shift," consider the transition from Aristotelian agent-centered virtue ethics to Kantian rule-centered obligation ethics.

21. Galileo, *Two New Sciences*, tr. S. Drake (Toronto: Wall & Thompson, 1989), 197 et passim.

22. Leibniz is exceptional in this regard too. Every corporeal substance is a machine whose parts are organized "all the way down" in virtue of the single dominant monad within it. No manmade machine is so organized. All material particulars having no dominant monad within them are not real unities; they are what I have called *pragmata*.

23. Aristotle prescribed no rules for investigating nature and would have rejected the claim that we can lay down such rules in advance. In fact his empirical investigations are not methodical in our sense of the term. Galilean investigations of nature, on the other hand, are plainly methodical, and several of its early proponents did lay down such rules (*regulae*) in advance—among them Bacon, Descartes, Spinoza, and Newton.

24. Briefly, he found that final causality had little application to elementary masses, that the formal causality of these masses was very difficult to describe, that efficient causality had little application to the heavenly spheres, and that the matter of these spheres had to be very different from the matter of sublunary things.

25. Bacon, *New Organon* 1.3–4, 58, 66, 98, 121, 124, 129, 2.4, 17. See also Descartes, *Discourse of Method* 2, *Meditations* 3, 6, and Hobbes, *Leviathan* 46.

26. See chapter 7 of this volume.

27. Having stipulated that an electron or a proton has an electric charge = ±1, one must assign fractional charges to quarks; but this does not affect the point being made here.

28. See chapter 6 of this volume.

29. See chapter 6 of this volume.

30. See chapter 6 of this volume.

31. Plato, *Symposium* 203b2–e5.

32. Parmenides, B.1.28–30 and B8.51f (DK).

33. Parmenides, B8.38–41, 52–4, 60, B19.1–3 (DK). Cf. S. Umphrey, "Not Is Not," in *The Envisioned Life: Essays in Honor of Eva Brann*, ed. P. Kalkavage and E. Salem (Philadelphia, PA: Paul Dry Books, 2007), 350–74.

34. Plato, *Phaedo* 99b4–100b7.

35. Plato, *Timaeus* 27d5–28a6, 59c5–d2.

36. Plato, *Timaeus* 17b and context, *Critias* 106a and context.

37. Epicurus, for example, argued that the observed division of things into discrete kinds could be explained only if their atomic constituents have finitely many shapes (*Letter to Herodotus*, 55–56). He seems, however, not to have offered any detailed of account of these atomic kinds or of the continuants that belong to them in virtue of their natures, nor does he seem to have explained just how these atomic kinds underwrite the heterogeneity we observe.

38. Cf., e.g., Aetius 46A. Stoics agreed that something exists if and only of it is corporeal; that all happenings accord with the *logos* and are thus fated; and that elementary masses, pneuma, human beings, and the present world-order (*kosmos*) as a whole have all emerged from primordial fire. Like Empedocles, however, and unlike Aristotle, they seem to have been little interested in determining the ontic status of these emergent things.

39. Thomas Aquinas exemplifies the former aim, William of Ockham, the latter. It should be noted that Ptolemaic astronomy also prevailed at this time. In the words of my late colleague, Curtis Wilson, they were both "doomed to succeed."

40. The fourteenth-century "Calculators" did find that some propositions belonging to Aristotle's mechanics (in *Physics* 4.8, 6.1–2, 7.45) are incorrect, and replaced them with propositions that anticipate developments in Galilean mechanics. But the framework within which they made these corrections remained Aristotelian. Cf. C. Wilson, *William Heytesbury: Medieval Logic and the Rise of Mathematical Physics* (Madison, WI: Wisconsin University Press, 1956), 24ff, 57, 115ff.

41. Cf., e.g., Kepler, *Epitome of Copernican Astronomy*, 4.4.2–4, 5.2.1; Galileo, *Two New Sciences*, 153ff; and E. Cassirer, *The Individual and the Cosmos in Renaissance Philosophy*, tr. M. Domandi (New York: Harper, 1964), 172–77; A. Koyré, *Galileo Studies*, tr. J. Nepham (Atlantic Highlands, NJ: Humanities Press, 1978), 37f, 74, 183,199, 208f.

42. Kant, *Critque of Pure Reason*, e.g., B17ff, A20–33/B34–49, A48/65, A106, A114, A126–28/B159–60, B163–65, A253–54/B309, B163–65, A421ff/B449ff. See also his *Prolegomenon to any Future Metaphysics*, 5–6, 9, 11, 16–17, 28.

43. An example can be found in Kant's *Metaphysical Foundations of Science*, where he sets himself the task of clarifying the empirical concept of matter in motion. Empirical concepts, as such, are proper to science and not philosophy. Kant's task is thus one of "reconstruction," he says, and not *a priori* construction or deduction.

# Bibliography

Abbott, Barbara. 1997. "A Note on the Nature of 'Water.'" *Mind* 106: 311–19.
Aitchison, I. J. R. 1985. "Nothing's Plenty: The Vacuum in Modern Quantum Field Theory." *Contemporary Physics* 26.4: 333–91.
———, and A. J. G. Hey. 2003–2004. *Gauge Theories in Particle Physics*, 3rd ed., volumes I and II. New York: Taylor & Francis.
Alberts, Bruce, A. Johnson, J. Lewis, et al. 2002. *Molecular Biology of the Cell*, 4th edition. New York: Garland Science.
Almog, Joseph. 1991. "The What and the How." *Journal of Philosophy* 88: 225–44.
Amundson, Ron, and G. V. Lauder. 1993–1994. "Function without Purpose." *Biology and Philosophy* 9: 443–69.
Armstrong, David M. 1978a. *Nominalism and Realism.* New York, Cambridge University Press.
———. 1978b. *A Theory of Universals.* New York: Cambridge University Press.
———. 1985. *What Is a Law of Nature?* New York: Cambridge University Press.
———. 1988. "Can a Naturalist Believe in Universals?" In *Science and Reflection*, ed. E. Ullmann-Margalit. Boston: Kluwer.
———. *Universals: An Opinionated Introduction.* 1989. Boulder, CO: Westview Press.
———. 1993. *A Materialist Theory of Mind.* New York: Routledge.
———. 1997. *A World of States of Affairs.* New York: Cambridge University Press.
Aristotle. 1936. *Categoriae et Liber de Interpretatione*, ed. L. Minio-Paluello. Oxford: Clarendon Press.
———. 1957. *Metaphysica*, ed. W. Jaeger. Oxford: Clarendon Press.
Atkins, Peter. 1991. *Quanta: a Handbook of Concepts.* New York: Oxford University Press.
———. 1994. *Physical Chemistry*, 5th ed. New York: Freeman.
Atran, Scott, and Douglas L. Medin. 2008. *The Native Mind and the Cultural Construction of Nature.* Cambridge, MA: MIT Press.
Auyang, Sunny Y. 1995. *How is Quantum Field Theory Possible?* New York: Oxford University Press.
Bacon, John. 1995. *Universals and Property Instances: The Alphabet of Being.* Oxford: Blackwell.
Baird, Davis and E. Scerri, L. McIntyre, ed. 2006. *Philosophy of Chemistry: Synthesis of a New Discipline.* Dordrecht: Springer.
Baker, Lynn R. 2013 *Naturalism and the First-Person Perspective.* New York: Oxford University Press.
Barker, Matthew. 2007. "The Empirical Inadequacy of Species Cohesion by Gene Flow." *Philosophy of Science* 74: 654–65.
———, and R. A. Wilson. 2010. "Cohesion, Gene Flow, and the Nature of Species." *Journal of Philosophy* 107: 59–77.
Bealer, George. 1982. *Quality and Concept.* Oxford: Clarendon Press.
———. 1987. "The Limits of Scientific Essentialism." In *Philosophical Perspectives I: Metaphysics*, ed. J. E.Tomlinson. Atascadero, CA: Ridgeview Publishing. 289–365.
———. 1998. "Intuition and Autonomy in Philosophy." In M. DePaul and W. Ramsey, 201–39.
Bedau, Mark A., and P. Humphreys, ed. 2008. *Emergence.* Cambridge, MA: MIT Press.
Beebee, Helen and N. Sabbarton-Leary, ed. 2010. *The Semantics and Metaphysics of Natural Kinds.* New York: Routledge.

Belousek, D. W. "Non-Separability, Non-Supervenience, and Quantum Ontology." 2003. *Philosophy of Science* 70: 791–811.
Bergmann, Gustav. 1954. *The Metaphysics of Logical Positivism.* New York: Longmans, Green, & Co.
———. 1967. *Realism: A Critique of Brentano and Meinong.* Madison, WI: Wisconsin University Press.
Bhushan, Nalini. "Are Chemical Kinds Natural Kinds?" 2006. In D. Baird, E. Scerri, and L. McIntyre, 328–36.
———, and S. Rosenfeld, ed. 2000. *Of Minds and Molecules: New Philosophical Perspectives on Chemistry.* New York: Oxford University Press.
Bigelow, John C. 1999. "Scientific Ellisianism." In *Causation and the Laws of Nature,* ed. H. Sankey. Dordrecht: Kluwer. 45–59.
———, B. Ellis, and C. Lierse. 1992. "The World as One of a Kind: Natural Necessity and Laws of Nature." *British Journal for the Philosophy of Science* 43: 371–88.
Bird, Alexander. 2007. *Nature's Metaphysics: Laws and Properties.* New York: Oxford University Press.
Blackman, Larry L., ed. 2005. *The Philosophy of Panayot Butchvarov.* Lewiston, NY: Edwin Mellen.
Borgesson, Gary. 1999. "Not For Their Own Sake: Species and the Riddle of Individuality." *Review of Metaphysics* 52: 867–96.
Boyd, Richard. 1991. "Realism, Anti-Foundationalism, and the Enthusiasm for Natural Kinds." *Philosophical Studies* 61: 127–48.
———. 1999. "Kinds, Complexity, and Multiple Realization." *Philosophical Studies* 95: 67–98.
———. 2010. "Realism, Natural Kinds, and Philosophical Methods." In H. Beebee and N. Sabbarton-Leary, 212–34.
Brakel, Jaap van. 1997. "Chemistry as the Science of the Transformation of Substances." *Synthese* 111: 253–82.
———. 2000. *Philosophy of Chemistry.* Leuven, Belgium: Leuven University Press.
———. 2000. "The Nature of Chemical Substances." In N. Bhushan and S. Rosenfeld, 162–84.
Brody, Baruch. 1980. *Identity and Essence.* Princeton, NJ: Princeton University Press.
Buss, Leo W. 1987. *The Evolution of Individuality.* Princeton, NJ: Princeton University Press.
Butchvarov, Panayot. 1966. *Resemblance and Identity.* Bloomington: Indiana University Press.
———. 1974. "The Limits of Ontological Analysis." In *The Ontological Turn,* ed. M. S. Gram and E. D. Klemke. Iowa City: Iowa University Press.
———. 1979. *Being Qua Being: A Theory of Identity, Existence and Predication.* Bloomington: Indiana University Press.
———. 1998. *Skepticism about the External World.* New York: Oxford University Press.
Campbell, Keith. 1990. *Abstract Particulars.* Oxford: Blackwell.
Cao, Tian U., ed. 2004. *Conceptual Foundations of Quantum Field Theory.* New York: Cambridge University Press.
Cartwright, Nancy. 1983. *How the Laws of Physics Lie.* New York: Oxford University Press.
Casey, Dylan.2013. "Fields and Particles and Being." *St. John's Review* 54: 57–88.
Cassirer, Ernst. 1956. *Determinism and Indeterminism in Modern Physics.* New Haven, CT: Yale University Press.
Chakravartty, Anjan. 2000. "The Structuralist Conception of Objects." *Philosophy of Science* 70: S867–S878.
Chen, Xiang. "Object and Event Concepts: A Cognitive Mechanism of Incommensurability." 2003. *Philosophy of Science* 70: S962–S974.
Coyne, Jerry A., and H. A. Orr. 2004. *Speciation.* Sunderland, MA: Sinauer.
Cummins, Robert. "Functional Analysis." 1975. *Journal of Philosophy* 72: 741–60.

Davies, Paul S. 2001. *Norms of Nature: Naturalism and the Nature of Functions.* Cambridge, MA: MIT Press.
Dennett, Daniel. 2013. "Kinds of Things—Towards a Bestiary of the Manifest Image." In *Scientific Metaphysics*, ed. D. Ross, J. Ladyman, and H. Kincaid. New York: Oxford University Press. 96–107.
DePaul, Michael, and W. Ramsey, ed. 1998. *Rethinking Intuition.* Lanham, MD: Rowman & Littlefield.
de Queiroz, Kevin. 1998. "The General Lineage Concept of Species, Species Criteria, and the Process of Speciation." In *Endless Forms*, ed. Daniel J. Howard and S. H. Berlocher. New York: Oxford University Press. 57–76.
———. 1999. "The General Lineage Concept of Species and the Defining Properties of the Species Category." In R. A Wilson, 49–89.
de Sousa, Ronald. 1984. "The Natural Shiftiness of Natural Kinds." *Canadian Journal of Philosophy* 14: 565–80.
Devitt, Michael. 1991. *Realism and Truth.* Oxford: Blackwell.
———. 2008. "Resurrecting Biological Essentialism." *Philosophy of Science* 75: 344–82.
Donellan, Keith. 1998. "Kripke and Natural Kind Terms." In *Knowledge and Mind: Philosophical Essays*, ed. C. Ginet and S. Shoemaker. New York: Oxford University Press. 84–104.
Dummett, Michael. 1978. *Truth and Other Enigmas.* Cambridge, MA: Harvard University Press.
Dupré, John. 1981. "Natural Kinds and Biological Taxa." *The Philosophical Review* 90: 66–90.
———. 1993. *The Disorder of Things: Metaphysical Foundations of the Disunity of Science.* Cambridge, MA: Harvard University Press.
———. 2002. "Is 'Natural Kind' a Natural Kind Term?" *Monist* 85: 29–49.
Earley, J. E. 2006. "Some Philosophical Implications of Chemical Symmetry." In D. Baird, E. Scerri, and L. McIntyre 2006, 207–20.
Ellis, Brian. 2001. *Scientific Essentialism.* New York: Cambridge University Press.
———. 2002. *The Philosophy of Nature: a Guide to the New Essentialism.* Montreal: McGill-Queen's University Press.
———. 2005. "Universals, the Essential Problem, and Categorical Properties." *Ratio* 18: 462–72.
———. 2010. *Metaphysics of Scientific Realism.* Montreal: McGill-Queens University Press.
Ereshefsky, Marc, ed. 1992. *The Units of Evolution.* Cambridge, MA: MIT Press.
———. 1998. "Species Pluralism and Antirealism." *Philosophy of Science* 65: 103–20.
———. 1999. "Species in the Linnaean Hierarchy." In R. A. Wilson, 285–305.
Fales, Evan. 1990. *Causation and Universals.* New York: Routledge.
———. 1996. *A Defense of the Given.* Lanham, MD: Rowman & Littlefield.
Fine, Kit. 1975. "How to Compare Theories: Reference and Change." *Nous* 9: 17–32.
Foley, Richard. 1998. "Rationality and Intellectual Self-Trust." In M. DePaul and W. Ramsey, 241–56.
Forbes, Graeme. 1985. *The Metaphysics of Modality.* New York: Oxford University Press.
French, Peter A., T. E. Uehling, Jr., and H. K. Wettstein, ed. 1988. *Midwest Studies in Philosophy XII: Realism and Antirealism.* Minneapolis: Minnesota University Press.
French, Steven. 1998. "On the Withering Away of Physical Objects." In *Interpreting Bodies: Classical and Quantum Objects in Modern Physics*, ed. E. Castellani. Princeton, NJ: Princeton University Press. 93–113.
———. 2015. "Identity and Individuality in Quantum Theory." In *The Stanford Encyclopedia of Philosophy*. http://plato.stanford.edu/entries/qt-idind/.
———, and D. Krause. 2006. *Identity in Physics: A Historical, Philosophical, and Formal Analysis.* Oxford: Clarendon Press.
Gelman, S. A., and L. A Hirschfeld. 1999. "How Biological Is Essentialism?" In D. Medin and S. Atran, ed., 403–46.
Ghiselin, Michael. 1997. *Metaphysics and the Origin of Species.* Albany, NY: SUNY Press.

Gilbert, Scott F. 1994. *Developmental Biology*, 4th ed. Sunderland, MA: Sinauer.
———, and D. Epel. 2009. *Ecological Developmental Biology: Integrating Epigenetics, Medicine, and Evolution.* Sunderland, MA: Sinauer.
Goodman, Nelson. 1978. *Ways of Worldmaking*. Indianapolis, IN: Hackett.
Greene, Brian. 1999. *The Elegant Universe.* New York: Norton.
Griffiths, David. 2008. *Introduction to Elementary Particles* 2nd ed. Weinheim: Wiley-VCH.
Gurwitsch, Aron. 1964. *The Field of Consciousness.* Pittsburgh: Duquesne University Press.
Hacking, Ian. 1988. *Representing and Intervening.* New York: Cambridge University Press.
———. 1991. "A Tradition of Natural Kinds." *Philosophical Studies* 61: 109–26.
———. 2007. "Natural Kinds, Rosy Dawn, Scholastic Twilight." *Royal Institute of Philosophy*, S82: 202–39.
Harré, Rom, and E. H. Madden. 1975. *Causal Powers: A Theory of Natural Necessity.* Totowa, NJ: Rowman & Littlefield.
Heller, Mark. 1988. "Vagueness of the Standard Ontology." *Nous* 22: 109–31.
———. 1990. *The Ontology of Physical Objects: Four-Dimensional Hunks of Matter.* New York: Cambridge University Press.
———. 1992. "Things Change." *Philosophy and Phenomenological Research* 52: 695–704.
Hirsch, Eli. 1976a. *The Persistence of Physical Objects.* Philadelphia: University City Science Center.
———. 1976b. "Physical Identity." *Philosophical Review* 67: 376–89.
Horgan, Terrence. 1994. "Vagueness and the Forced-March Sorites Paradox." In *Philosophical Perspectives 8: Logic and Language*, 159–88.
Howard, Daniel J., and S. Berlocher, ed. 1998. *Endless Forms: Species and Speciation.* New York: Oxford University Press.
Hull, David. L. 1978. "A Matter of Individuality." *Philosophy of Science* 45: 335–60. Reprinted in E. Sober 1994.
———. 1980. "Individuality and Selection." *Annual Review of Ecology and Systematics* 11: 311–32.
———. 1984. "Historical Entities and Historical Narratives." In *Minds, Machines and Evolution*, ed. C. Hookway. New York: Cambridge University Press.
———. 1989. *The Metaphysics of Evolution.* Albany, NY: SUNY Press.
———. 1992. "Biological Species: An Inductivist's Nightmare." In *How Classification Works*, ed. M. Douglas and D. L. Hull. Edinburgh: Edinburgh University Press.
———. 1998. "Individual." In *Keywords in Evolutionary Biology*, ed. Evelyn F. Keller and E. A. Lloyd. Cambridge, MA: Harvard University Press.
Hume, D. 2002. *Treatise of Human Nature*, ed. D.F. and M. J. Norton. NEW YORK: Oxford University Press.
Husserl, Edmund. 1962. *Ideas I: General Introduction to Phenomenology*, tr. W. R. B. Gibson. New York: Collier Books.
———. 1970. *Logical Investigations*, two vols., tr. J. N. Findlay. London: Routledge & Kegan Paul.
———. 1973. *Experience and Judgment.* Edited L. Landgrabe, trans. J. S. Churchill and K. Ameriks. Evanston: Northwestern University Press.
Inwagen, Peter van. 1990. *Material Beings.* Ithaca, NY: Cornell University Press.
Jenkins, Zack. 2003. "Do You Believe in Orbitals?" *Philosophy of Science* 70: S1053–S1057.
Johnston, Mark.1997. "Manifest Kinds." *Journal of Philosophy* 94: 564–83.
Kant, I. 1996. *Critique of Pure Reason*, tr. W. S. Pluhar. Indianapolis, Hackett Publishing.
Keilin, D. 1959. "The Problem of Anabiotic or Latent Life." *Proceedings of the Royal Society of London*," ser. B 150: 149–91.
Kitcher, Philip. 1992. "The Naturalists Return." *The Philosophical Review* 101: 53–114.

———. 1993. "Function and Design." In *Midwest Studies in Philosophy: Philosophy of Science*, vol. 18, ed. P. A. French, T. E. Uehling, and H. K. Wettstein. Notre Dame, IN: Notre Dame University Press. 379–97.

———. 2001. "Real Realism: The Galilean Strategy." *The Philosophical Review* 110: 151–97.

Khalidi, Muhammad A. 2009. "How Scientific Is Scientific Essentialism?" *Journal for General Philosophy of Science* 40: 85–101.

———. 2015. *Natural Categories and Human Kinds: Classification in the Natural and Social Sciences*. New York: Cambridge University Press.

Khlentzos, Drew. 2004. *Naturalistic Realism and the Antirealist Challenge*. Cambridge, MA: MIT Press.

Kornblith, Hilary. 2013. "Naturalism vs. the First-Person Perspective." *Proceedings & Addresses of the American Philosophical Association*, 87: 122–41.

Kripke, Saul. 1980. *Naming and Necessity*. Oxford: Blackwell.

Kuhn, Thomas. 1970. *The Structure of Scientific Revolutions*, 2nd ed. Chicago: Chicago University Press.

———. 1977. *The Essential Tension*. Chicago: Chicago University Press.

———. 1977a. "Second Thoughts on Paradigms." In Frederick Suppes, *The Stucture of Scientific Theories*. Urbana: Illinois University Press. 459–82.

Ladyman, James and D. Ross. 2007. *Every Thing Must Go: Metaphysics Naturalized*. New York: Oxford University Press.

Lakoff, George. 1987. *Women, Fire, and Dangerous Things: What Categories Reveal about the Mind*. Chicago: Chicago University Press.

Landesman, Charles, ed. 1971. *The Problem of Universals*. New York: Basic Books.

LaPorte, Joseph. 2004. *Natural Kinds and Conceptual Change*. New York: Cambridge University Press.

Laycock, Henry. 1972. "Some Questions of Ontology." *Philosophical Review* 81: 3–42.

———. 2006. *Words without Objects*. Oxford: Clarendon Press.

Lewontin, Richard. 1970. "The Units of Selection." *Annual Review of Systematics and Ecology* 1: 1–18.

———. 1983. "The Organism as the Subject and Object of Evolution." *Scientia* 118: 63–82.

———. 2000. *The Triple Helix*. Cambridge, MA: Harvard University Press.

Locke, John. 1979. *An Essay Concerning Human Understanding*, ed. P. H. Nidditch. New York: Oxford University Press.

Loux, Michael J., ed. 1970. *Universals and Particulars*. Garden City, NY: Doubleday.

———. 1974. "Kinds and the Dilemma of Individuation." *Review of Metaphysics* 27: 773–84.

———. 1976. "The Concept of a Kind." *Philosophical Studies* 29: 53–61.

———. 1978. *Substance and Attribute*. Dordrecht: Reidel.

Lowe, E. J. 2006. *The Four-Category Ontology: A Metaphysical Foundation for Natural Science*. New York: Oxford University Press.

Mackie, Penelope. 2006. *How Things Might Have Been*. Oxford: Clarendon Press.

Magnus, P. D. 2012. *Scientific Enquiry and Natural Kinds: From Planets to Mallards*. New York: Palgrave Macmillan.

Martin, C. B. 1980. "Substance Substantiated." *Australasian Journal of Philosophy* 58: 3–10.

Maynard Smith, John. 2000. "The Concept of Information in Biology." *Philosophy of Science* 67: 177–94.

———, and E. Szathmáry. 1995. *The Major Transitions in Evolution*. New York: Oxford University Press.

Mayr, Ernst. 1942. *Systematics and the Origin of Species*. New York: Columbia University Press.

———. 1961. "Cause and Effect in Biology." *Science* n.s. 134: 1501–06.

———. 1982. *The Growth of Biological Thought: Diversity, Evolution, and Inheritance*. Cambridge, MA: Harvard University Press.

———. 1988. *Toward a New Philosophy of Biology: Observations of an Evolutionist.* Cambridge, MA: Harvard University Press.
McGinn, Colin. 1976. "On the Necessity of Origin." *Journal of Philosophy* 73: 127–35.
Medin, Douglas L., and S. Atran, ed. 1999. *Folkbiology*. Cambridge, MA: MIT Press.
Mellor, D. H. 1977. "Natural Kinds." *British Journal for the Philosophy of Science* 28: 299–312.
Mill, John S. 1973–1974. *A System of Logic*, ed. J. M. Robson. Toronto: Toronto University Press.
Milliken, Ruth G. 1989. "In Defense of Proper Functions." *Philosophy of Science* 56: 288–302.
Moreland, James P. 1985. *Universals, Qualities, and Quality-Instances.* Lanham, MD: University Press of America.
———. 2001. *Universals.* Montreal: McGill University Press.
Mumford, Stephen. 2004. *Laws in Nature.* New York: Routledge.
Mynott, Jeremy. 2009. *Birdscapes.* Princeton, NJ: Princeton University Press.
Neander, Karen. 1991. "Functions as Selected Effects: The Conceptual Analyst's Defense." *Philosophy of Science* 58: 168–84.
Needham, Paul. 2000. "What is Water?" *Analysis* 60: 13–21.
———. 2006. "Aristotle's Theory of Chemical Reaction and Chemical Substances." In Baird, Scerri, and McIntyre, 43–68.
———. 2006a. "Substance and Modality." *Philosophy of Science* 73: S829–S840.
Newman, Andrew. 1992. *The Physical Basis of Predication.* New York: Cambridge University Press.
Noonan, H. 1983. "The Necessity of Origin." *Mind* 92: 1–20.
Okasha, Samir. 2002. "Darwinian Metaphysics: Species and the Question of Essentialism." *Synthese* 131: 191–213.
Pauling, Linus. 1970. *General Chemistry.* New York: Dover.
Peirce, Charles. S. 1940. "Kind." In *Dictionary of Philosophy and Psychology* I. Edited J. M. Baldwin. New York: Peter Smith.
Plantinga, Alvin. 1974. *The Nature of Necessity.* New York: Oxford University Press.
Plato. 1962. *Platonis Opera*, ed. J. Burnett, Oxford: Clarendon Press.
Poncinie, Larry. 1985. "Meaning Change for Natural Kind Terms." *Nous* 19: 415–27.
Preston, Beth. 1998. "Why is a Wing Like a Spoon?" *Journal of Philosophy* 95: 215–54.
Price, H. H. 1953. *Thinking and Experience.* Cambridge, MA: Harvard University Press.
Psillos, Stathis. 2002. *Causation and Explanation*. Montreal: McGill-Queens.
Putnam, Hilary. 1978. *Meaning and the Moral Sciences.* Boston: Routledge and Kegan Paul.
———. 1979. *Mind, Language and Reality: Philosophical Papers*, vol. 2. New York: Cambridge University Press.
———. 1981. *Reason, Truth, and History.* New York: Cambridge University Press.
———. 1983. *Realism and Reason: Philosophical Papers*, vol. 3. New York: Cambridge University Press.
———. 1987. *The Many Faces of Realism.* LaSalle, IL: Open Court.
———. 1989. *Representation and Reality.* Cambridge, MA: MIT Press.
———. 1990. *Realism with a Human Face.* Cambridge, MA: Harvard University Press.
———. 1995. *Renewing Philosophy*. Cambridge, Ma: Harvard University Press.
Quine, Willard. Van O. 1960 *Word and Object*. Cambridge, MA: MIT Press.
———. 1969. "Natural Kinds." In *Ontological Relativity and Other Essays.* New York: Columbia University Press.
———. 1976. "Whither Physical Objects?" In *Essays in Memory of Imre Lakatos*, ed. R. S. Cohen, P. K. Feyerabend, and M. W. Wartofsky. Dordrecht: Reidel.
Ramsey, Jeffrey L. 1997. "Molecular Shape, Reduction, Explanation and Approximate Concepts." *Synthese* 111: 233–51.
———. 2000. "Realism, Essentialism, and Intrinsic Properties: A Case for Molecular Shape." In N. Bhushan and S. Rosenfeld, 117–28.
Robinson, Howard. 1982. *Matter and Sense.* New York: Cambridge University Press.

Rosenberg, Alexander. 1985. *The Structure of Biological Science.* New York: Cambridge University Press.
———, and K. Neander. 2009. "Are Homologies (Selected Effect or Causal Role) Function Free?" *Philosophy of Science* 76: 307–34.
Ruse, Michael. 1987. "Biological Species: Natural Kinds, Individuals, or What?" *British Journal for the Philosophy of Science* 38: 225–42.
Russell, Bertrand. 1919. "On the Notion of Cause." In B. Russell, *Mysticism and Logic.* New York: Longmans, Green and Co.
———. 1940. *An Inquiry into Meaning and Truth.* London: Allen & Unwin.
———. 1948. *Human Knowledge, Its Scope and Limits.* London: Allen & Unwin.
———. 1975. *The Problems of Philosophy.* New York: Oxford University Press.
Salmon, Nathan U. 1981. *Reference and Essence.* Princeton, NJ: Princeton University Press.
Scerri, Eric R. 1994. "Has Chemistry Been At Least Approximately Reduced to Quantum Mechanics?" *Philosophy of Science* 61: 160–70.
———. 2005. "Some Aspects of the Metaphysics of Chemistry and the Nature of the Elements." *Hyle* 11: 127–45.
———. 2007. *The Periodic Table: Its Story and Significance.* New York: Oxford University Press.
Schütz, Alfred. 1967a. *Collected Papers*, vol. 1. The Hague: Nijhoff.
———. 1967b. *The Phenomenology of the Social World.* Trans. G. Walsh and F. Lehnert. Evanston, IL: Northwestern University Press.
Schwarz, Stephen P. 1978. "Putnam on Artifacts." *Philosophical Review* 87: 566–74.
———. 1980. "Natural Kinds and Nominal Kinds." *Mind* n.s.89: 182–95.
Sider, Theodore. 1997. "Four-Dimensionalism." *Philosophical Review* 106: 197–231.
Simons, Peter. 1994. "Particulars in Particular Clothing: Three Trope Theories of Substance." *Philosophy and Phenomenological Research* 54: 553–75.
Slater, Matthew H. 2005. "Monism on the One Hand, Pluralism on the Other." *Philosophy of Science* 72: 22–42.
Smolin, Lee. 1997. *The Life of the Cosmos.* New York: Oxford University Press.
Sober, Elliott. 1980. "Evolution, Population Thinking, and Essentialism." *Philosophy of Science* 47: 350–83.
———. 1982. "Why Logically Equivalent Predicates Pick Out Different Properties." *American Philosophical Quarterly* 19: 183–89.
———. 1993. *Philosophy of Biology.* Boulder, CO: Westview Press.
———, ed. 1994. *Conceptual Issues in Evolutionary Biology*, 2nd edition. Cambridge, MA: MIT Press.
Sterelny, Kim, and P. E. Griffiths. 1999. *Sex and Death: An Introduction to Philosophy of Biology.* Chicago: Chicago University Press.
Strawson, Peter. 1959. *Individuals: An Essay in Descriptive Metaphysics.* London: Methuen.
———. 1979. "Universals." In *Midwest Studies in Philosophy IV: Studies in Metaphysics*, ed. P. A. French, T. E. Uehling, Jr., and H. K. Wettstein. Minneapolis: Minnesota University Press.
Swoyer, Chris. 1982. "The Nature of Natural Laws." *Australasian Journal of Philosophy* 60: 203–23.
———. 2005. "Properties," *Stanford Encyclopedia of Philosophy.* http://plato.stanford.edu/entries/properties/.
Teller, Paul. 1986. "Relational Holism and Quantum Mechanics." *British Journal for the Philosophy of Science* 37: 71–81.
———. 1990. "Prolegomenon to a Proper Interpretation of Quantum Field Theory." *Philosophy of Science* 57: 594–618.
Timmermans, Jean. 1963. *The Concept of Species in Chemistry*, tr. R. E. Oesper. New York: Chemical Publishing Co.
Umphrey, Stewart. 2002. *Complexity and Analysis.* Lanham, MD: Lexington Books.
Unger, Peter. 1979. "There Are No Ordinary Things." *Synthese* 41: 117–54.

Venn, John. 1923. *The Logic of Chance: an Essay on the Foundations and Province of the Logic of Probability*. New York: Chelsea.

Vermeeren, Hans P. W. 1986. "Controversies and Existence Claims in Chemistry: The Theory of Resonance." *Synthese* 69: 273–90.

Wagner, Steven J., and R. Warner, ed. 1993. *Naturalism: A Critical Appraisal*. Notre Dame, IN: Notre Dame University Press.

Weinberg, Steven. 1977. "Notes for a History of Quantum Field Theory." *Daedalus* 106: 17–35.

———. 1992. *Dreams of a Final Theory*. New York: Pantheon.

———. 1995. "Nature Itself." In *Twentieth Century Physics*, ed. L. Brown, B. Pippard, and A. Pais. New York: Taylor & Francis. 2032–40.

———. 2004. "What is Quantum Field Theory, and What Did We Think it Was?" In T. U. Cao, 241–51.

Weisberg, Michael. 2006. "Water is *Not* $H_2O$." In D. Baird, E. Scerri, L. McIntyre, 337–45.

Wiggins, David. 1980. *Sameness and Substance*. Cambridge, MA: Harvard University Press.

———. 2001. *Sameness and Substance Renewed*. New York: Cambridge University Press.

Wigner, Eugene P. 1967. *Symmetries and Reflections*. Bloomington, IN: Indiana University Press.

Wilkerson, T. E. 1995. *Natural Kinds*. Aldershot: Avebury.

Williams, D. C. 1953. "The Elements of Being." *Review of Metaphysics* 7: 3–18, 171–92.

Williamson, Timothy. 1994. *Vagueness*. New York: Routledge.

Wilson, Jack. 1999. *Biological Individuality: The Identity and Persistence of Living Entities*. New York: Cambridge University Press.

Wilson, Robert A., ed. 1999. *Species: New Interdisciplinary Essays*. Cambridge, MA: MIT Press.

Wittgenstein, Ludwig. 1968. *Philosophical Investigations*, 3rd ed., tr. G. E. M. Anscombe. New York: MacMillan.

Woody, Andrea I. 2000. "Putting Quantum Mechanics to Work in Chemistry." *Philosophy of Science* 67: S612–S627.

Wright, C., 2000. "Truth as Sort of Epistemic: Putnam's Peregrinations," *Journal of Philosophy* 97.6: 335–64.

Wright, Larry.1973. "Functions." *Philosophical Review* 82: 139–68.

Zalta, Edward N. 1983. *Abstract Objects: an Introduction to Axiomatic Metaphysics*. Dordrecht: Reidel.

Zemach, Edward M. 1977. "Putnam's Theory on the Reference of Substance Terms." *Journal of Philosophy* 73: 116–27.

Zimmerman, Dean W.1995. "Theories of Masses and Problems of Constitution." *The Philosophical Review* 104: 53–110.

# Index

# About the Author

**Stewart Umphrey** is Tutor Emeritus at St. John's College, Annapolis, Maryland. He formerly held positions in departments of philosophy at SUNY-Geneseo, The Graduate Faculty of the New School for Social Research, and the University of Oklahoma. His publications include *Zetetic Skepticism, Complexity and Analysis*, and articles on topics in metaphysics, political philosophy, and ancient Greek philosophy.

www.ingramcontent.com/pod-product-compliance
Ingram Content Group UK Ltd.
Pitfield, Milton Keynes, MK11 3LW, UK
UKHW041123071026
15587UKWH00009B/126

9 781498 531412